CAMBRIDGE LIBRARY COLLECTION

Books of enduring scholarly value

Life Sciences

Until the nineteenth century, the various subjects now known as the life sciences were regarded either as arcane studies which had little impact on ordinary daily life, or as a genteel hobby for the leisured classes. The increasing academic rigour and systematisation brought to the study of botany, zoology and other disciplines, and their adoption in university curricula, are reflected in the books reissued in this series.

Memorials of Sir C.J.F. Bunbury

Sir Charles James Fox Bunbury (1809–86), the distinguished botanist and geologist, corresponded regularly with Lyell, Horner, Darwin and Hooker among others, and helped them in identifying botanical fossils. He was active in the scientific societies of his time, becoming a Fellow of the Royal Society in 1851. This nine-volume edition of his letters and diaries was published privately by his wife Frances Horner and her sister Katherine Lyell between 1890 and 1893. His copious journal and letters give an unparalleled view of the scientific and cultural society of Victorian England, and of the impact of Darwin's theories on his contemporaries. Volume 4 covers the years 1857–64. Bunbury correctly foresaw in October 1859 that 'Darwin's forthcoming book on Species ... is likely to cause no little combustion in the scientific world'. He provides a valuable commentary on its reception over the following months.

T0188092

Cambridge University Press has long been a pioneer in the reissuing of out-of-print titles from its own backlist, producing digital reprints of books that are still sought after by scholars and students but could not be reprinted economically using traditional technology. The Cambridge Library Collection extends this activity to a wider range of books which are still of importance to researchers and professionals, either for the source material they contain, or as landmarks in the history of their academic discipline.

Drawing from the world-renowned collections in the Cambridge University Library, and guided by the advice of experts in each subject area, Cambridge University Press is using state-of-the-art scanning machines in its own Printing House to capture the content of each book selected for inclusion. The files are processed to give a consistently clear, crisp image, and the books finished to the high quality standard for which the Press is recognised around the world. The latest print-on-demand technology ensures that the books will remain available indefinitely, and that orders for single or multiple copies can quickly be supplied.

The Cambridge Library Collection will bring back to life books of enduring scholarly value (including out-of-copyright works originally issued by other publishers) across a wide range of disciplines in the humanities and social sciences and in science and technology.

Memorials of Sir C.J.F. Bunbury

Volume 4: Middle Life Part 3

Edited by
Frances Horner Bunbury
and Katharine Horner Lyell

CAMBRIDGE
UNIVERSITY PRESS

CAMBRIDGE UNIVERSITY PRESS

Cambridge, New York, Melbourne, Madrid, Cape Town,
Singapore, São Paolo, Delhi, Tokyo, Mexico City

Published in the United States of America by Cambridge University Press, New York

www.cambridge.org
Information on this title: www.cambridge.org/9781108041157

© in this compilation Cambridge University Press 2011

This edition first published 1891
This digitally printed version 2011

ISBN 978-1-108-04115-7 Paperback

MEMORIALS

OF

Sir C. J. F. Bunbury, Bart.

EDITED BY HIS WIFE.

THE SCIENTIFIC PARTS OF THE WORK REVISED BY
HER SISTER, MRS. LYELL.

MIDDLE LIFE.

Vol. III.

MILDENHALL:
PRINTED BY S. R. SIMPSON, MILL STREET.
MDCCCXCI.

JOURNAL.

Mildenhall.

We have had Charles and Mary Lyell and Joanna Horner with us, from the 22nd to the 31st of January. I always find their visits too short. After my own father and my brothers Edward and Henry, there is no man whom I love and value equally with Charles Lyell.

We had also a visit (from Thursday to Saturday) from Professor and Mrs. Henslow and their youngest daughter. I had not seen Henslow (I think) since the summer of 1854. He is a little older in looks, but as lively, cheerful, and active as ever. His activity and versatility of mind are indeed wonderful; and living as he does, in an obscure out-of-the-way country parish, his attention is awake to everything that is going on. He has a powerful as well as extremely active mind, (deficient however in the imaginative or poetical element), and if he had devoted it in earnest to science, he might no doubt have made great advances; yet I think on the whole his talents are more essentially practical than scientific. Certainly he has now

1857. devoted his time and his talents so absolutely to the improvement of his parish, and to teaching there and at Ipswich, that the pursuit of pure science has fallen quite into the background with him. He is one of those men of great ability (and there are not a few such in our time) who devote themselves rather to the spread than the advancement of science.

Lyell, on the other hand, though zealous in the cause of education and of general improvement, is pre-eminently a man of science, and to my thinking, a true philosopher. It is delightful to see his eagerness about every new scientific discovery and every subject of scientific research ; and this not merely in relation to his own especial pursuit of geology, but to all branches of natural history. I had many interesting scientific talks with him during this visit, and I have put down in another book some notes of what I learned from him in this way ; but no mere notes of facts or opinions can do justice to the amount of scientific and philosophical instruction that I gain (or ought to gain) from his conversation.

Mr. Donne came to us on Friday, the 30th, but left us the next morning. On the Friday evening he gave a very good and interesting lecture to our Institute, on the Life and Poetry of Crabbe. I was glad to hear justice done to that poet, of whom I formerly read much, and with great pleasure, but who in this Wordsworthizing generation is much neglected. Mr. Donne is a man of whose society and conversation I am very fond.

Mildenhall,
4th February, 1857.

My Dear Mary,

Very many thanks for your kind letter, and good wishes, and for your beautiful and useful present of slippers, which are particularly acceptable. I was very, very sorry to part with you *(you* in the dual number). I had enjoyed your visit very much indeed, but I am very apt to feel, after parting with dear friends (and especially such as I cannot meet very often) that I have not had half the enjoyment of their company that I ought. But I never am in company with Charles Lyell without learning something.

This has been a fine bright day, but cold; much snow still remaining. We took a walk together in the garden, and have since been looking over some of my dried Ferns. We threw a quantity of bread on the gravel walk for the birds, and it very soon disappeared.

You seemed to have had a very pleasant party at Lord Granville's, but it must indeed have required a good deal of courage to go out to an evening party in such weather.

The Queen's speech appears even more barren than usual. No mention of peace with Persia; I am afraid we are in for a serious war there.

Pretty good accounts from Barton.

Pray give my love to all the party at number 17, and many thanks for their presents; especially my love to Joanna, whom I was very sorry indeed to

1857.   lose so soon.   I hope you will all come again in
finer and warmer weather.

Ever your very affectionate Brother,

C. J. F. BUNBURY.

From his Father.

Barton,

February 4th, 1857.

Very many happy returns of this day to you my
dear Charles.   May your life be long and prosperous, and may the powers of your mind remain
unimpaired.   Could not Fanny and you come
here next Saturday instead of Monday ?   Both
Emily and I should be very glad to have you sooner.

I had much pleasure in receiving a kind letter
from Lyell yesterday ; and I hope to answer it
this evening.

Ever most affectionately yours,

H. E. B.

Mildenhall,

February 5, 1857.

My dear Mr. Horner,

Very truly and heartily do I thank you for
your kind letter and good wishes.   Many as are the
blessings I have to be thankful for—and I gratefully
feel that I have been loaded with far more blessings
than ever I could deserve—you are quite right in
supposing that I prize as the chief of all, and as the
great comfort of my life, the company and affection
of the best of wives, and I trust I may never see a
birthday without her by my side.

It grieves me very much that dear Mrs. Horner 1857. should still continue to suffer so much from the cruel neuralgic complaint, which has so long tormented her. It is a comfort that her strength keeps up so remarkably, and that her appearance is so little changed. If it be good for her to travel, and she be inclined to try Mildenhall for a change of air and scene, I hope I need not say that we shall be delighted to see her here, and will do all we can to make her comfortable.

With much love to Mrs. Horner, and the rest of your family party, I am ever

Your affectionate Son-in-law,

C. J. F. BUNBURY.

Mildenhall,
April 2. 1857.

My dear Susan,

Many thanks for your very agreeable letter, and I am sorry that, amidst the occupations of preparing for our journey to Paris, I have hardly time to answer it as fully as it deserves. We returned from Barton the day before yesterday, and yesterday the boys' school was inspected, and the whole affair went off satisfactorily; the school is certainly making progress, and I am convinced is doing much good. Fanny's zeal and perseverance in the good cause have not been thrown away. I am happy to be able to agree with you in most of the opinions you have expressed about the elections; which, to say the truth, I did not expect; I thought you would have been indignant at the strong and

1857. general demonstration of feeling in favour of Lord Palmerston's policy.

I am glad that the Russianizing leaders of the Manchester League are thrown out, and glad too that Lord John has come in; for though I think he made a fatal mistake in his Vienna negotiations, I believe him to be an honest and conscientious politician, and he has certainly done great services in his time. The result of the elections, as far as they have gone yet, is certainly remarkable, and a strong testimony to the popularity of Lord Palmerston and his policy. I admit that he might be *too* strong; for though my opinion of him (as a Minister) is different from yours, I believe that he requires watching; but I expect there will be Radicals enough in the House to keep him in order.

I am much interested by your account of Gualterio, and glad that he has led you to think more charitably of poor Carlo Alberto. The career of Piedmont as a constitutional State, under such difficulties, has been most honourable to its people.

For my part, I have been reading nothing lately (scientific matters excepted) but Sir Charles Napier's* life, of which I have pretty nearly finished the second volume; deeply interesting I find it. He was a noble-minded man, and the extracts from his journals and letters show his beautiful character in the clearest light; one sees him just as he was, so characteristically exhibited, that I can often fancy I *hear* him say the things that I read there.

* By his brother General Sir William Napier.

I hope you are not working too hard at your 1857.
translation.*

With love, ever your affectionate brother,

C. J. F. BUNBURY.

JOURNAL.

Paris,
April 20th, 1857.

A most beautiful and very hot day.

Arrived at Mrs. Power's house, 11, Rue de
Monceaux, about 10 a.m., having left Amiens at 6.
After breakfast, walked about Paris for some hours,
admiring and gaining general ideas. The Tuileries
gardens are delightful now, with their Horse-chesnut
trees in fresh young leaf, and the charming groups
of children at play under the trees. The easterly
continuation of the Rue de Rivoli, executed since
1850, very fine.

April 21st.

To the Jardin des Plantes, delivered my letters
for the professors, but found none of them at home.
At the Jardin, the system of exclusion of the public
from the hot-houses, and from many parts even of
the open air collection, still kept up. Medicinal
and other useful plants cultivated in a compartment
open to the public. Avenue of Judas trees, of
considerable size, now covered with flower buds,

* Of Colletta's History of Naples.

1857.   Returned on foot, very leisurely, along the quays, amusing myself with the old print and book stalls.

The river views of Paris, superior, I think, to any other town scenery that I know.   The new buildings of the Louvre, the Place du Carrousel, &c., very noble.

Mr. and Mrs. Byrne dined here.

April 22nd.

Lounged along the Rue de Rivoli, &c.   Drove out with Fanny, saw M. and Madame Tourgueneff and Madam de Vigogne.

April 23rd.

To the Louvre, and went through many of the galleries of the museum, but only in a hasty way. The whole has been re-arranged since I was last at Paris, and is very magnificent.   The finest pictures of all the schools are in one great square saloon. What interested me most was the very extensive and instructive collection of original drawings and sketches by old masters, arranged on screens and on the walls of a long suite of rooms.

The day very cold and disagreeable.

A pleasant, small evening party at Mrs. Byrne's; Count Alfred de Vigny, a man of very pleasing manners, with whom Fanny had much talk; Mr. William Prescott, son of the historian, with his young wife, a very lively and agreeable person; M. Laugel, a young geologist, with his wife, who is an American, very young, very pretty and lively. Madame Tourgueneff and her daughter; Mrs.

Champion, widow of Col. Champion, the botanist,
who died of wounds received at Inkerman : General
Ulloa, an Italian refugee ; Colonel Danner ; M. de
Luca, a very scientific chemist, an Italian by origin,
but resident at Paris ; and some others.

———————

Went with Fanny to the Jardin des Plantes, by
appointment, to see M. Adolphe Brongniart, who
was extremely courteous and attentive to us. He
accompanied us through the magnificent galleries of
mineralogy and geology, and appointed another day
to show me the collections more in detail. In the
first gallery, on the ground floor, is the collection of
minerals, which seems to be beautifully arranged,
and that of rock specimens ; and above, the great
collection of fossil bones ; there are also, at the end
of this gallery, some large and fine paintings of
geological, particularly volcanic, phenomena. Next
comes a sort of hall, lofty but not large, with the
statue of Jussieu in the centre, and round the walls
tall and remarkable stems of Palms and tree Ferns ;
among them, a noble stem of Cyathea glauca from
the Isle of Bourbon ; the specimen of Alsophila
Perottetiana with a forked stem, figured in Adr.
de Jussieu's elementary work ; and a Borassus
remarkable for its trunk, irregularly branched
towards the top. The great gallery beyond this
contains the superb collections of fossil plants, of
recent woods, and of recent fruits and seeds, the
more succulent kinds preserved in spirits, the rest

1857. dry. M. Brongniart observed that it would require
a long and elaborate comparative study of the
structure of recent woods, before one could under-
take to determine the fossil ones with any degree of
accuracy ; that some families, such as the Proteaceæ
do seem to be characterized by a peculiar structure
of the wood, which may be recognized ; but that in
general, the present state of our knowledge does not
allow us to determine the affinities of dicotyledons
from the wood alone.

He agrees with me in thinking the Germans
rather overbold in their determinations of fossil
dicotyledonous leaves. He told us that the
Araucaria excelsa is hardy in the south of France,
and even fructifies there ; that the Salisburia ripens
fruit in the south of France, and he thinks that it
possibly might even at Paris, but though they have
a very large and old tree of it in the Jardin, it is
a male, and of the female they have only young
plants which have not yet flowered.

April 27th.

Weather abominably cold. We went to the exhibition of Paul Delaroche's pictures at the Ecole
des Beaux Arts ; much pleased with them :—
" Strafford going to Execution ; " " The Virgin at
the foot of the Cross " (reminding one of Guido's
style, and equal in expression to almost any Italian
picture I remember) ; the " Execution of Lady Jane
Grey ; " a " Portrait of Napoleon ; " " Napoleon
Crossing the Alps " (this I saw a few years ago at
the French Exhibition in Pall Mall) ; " Cardinal

Richelieu"—in his last illness, ascending the Rhone   1857.
in his State barge, and dragging Cinq Mars and De
Thou in his train to execution; " Mazarin ill in bed,
playing at cards, surrounded by a gay circle of ladies
and courtiers " (these last two pictures are beautiful
in colouring as well as excellent in composition) ;
" The Last Adieus of the Girondins;" and some
small pictures of sacred subjects : all these I admire
very much.   The " Death of Queen Elizabeth,"
and some others, do not please me.

We then called on M. Barrande, the great
Silurian geologist, who was very agreeable.   He is
superintending the publication of the second
volume of his great work, and showed us
a number of the beautiful plates, illustrating the
genera Orthoceras and Cyrtoceras ; also many of
those of Trilobites in the first volume, pointing out
in particular the very interesting and curious series
of changes through which he has traced them, from
the almost microscopic embryo to the most perfect
state of the animal.   He showed us figures of the
most ancient Trilobites, found in the *primordial* or
oldest fossiliferous rocks of Bohemia, where they are
accompanied, he says, not by Lingulæ, as in the
oldest fossiliferous rocks of Wales, but by species
of Orthis.   The absence of Lingula from the primordial rocks of Bohemia, while it characterizes
those of Britain, Scandinavia, and North America,
is remarkable.   M. Barrande talked of the enigma
of the anthracitic schists of the Alps, remarking that
it was a case as yet unique in geology, and a puzzle
as yet unsolved ; but he seems inclined to attach

1857. more importance to the *liassic* character of the
animals—ammonites and belemnites—than to the
carboniferous character of the plants. He does not
think it can be explained by any folding or overturn
of the beds, as the schists with plants and the lime-
stones with shells appear to *alternate repeatedly.* At
the same time, the plants and the shells are never
found actually in the same beds.

April 28.

To the Louvre, with Fanny, but we went late,
and had but little time; admired particularly
Raphael's charming "Vierge au Diademe;" also
another Holy Family by him ("La Belle Jardinière;")
also Titian's "Mistress," and the grand "Paolo
Veronese." Then walked along the Rue de Rivoli,
entertained by the brilliant shops; then to Galig-
nani's; and so home.

An evening party at Madame Tourgueneff's. M.
Tourgueneff's melancholy accounts of the social
state of Russia, the miserable effects of the system
of slavery. English literature, he says, very popular
in Russia, and gaining ground more and more; all
Dickens' works translated into Russian, and much
read, to be found in almost every town. Graver
works also, such as Macaulay and Grote, much read,
though prohibited. Brilliant and beautiful effect
of the lights in the Place de la Concorde, and along
the quays at night.

To the Jardin des Plantes ; spent two hours very agreeably with M. Brongniart, who was most attentive and pleasant. He showed me the beautiful rooms containing the immense herbaria, and explained the plan of arrangement ; showed me a part of Tournefort's herbarium, which is kept separate, and of the general herbarium ; of which Vaillant's was the foundation ; parts of the collection of fossil plants ; and some of the hothouses, which are rich, but certainly inferior to Kew. The Museum has lately received a great acquisition, the entire herbarium of the Jussieus ;* that part of it which belonged to the great Ant. Laur. de Jussieu is kept in its original state as left by him ; but Adrien de Jussieu added very largely to the collection. Thence to the Luxembourg, and saw the gallery of modern pictures, but was rather disappointed with it. Lounged along the quays, looking at prints, etc., and so home. Weather still very cold and disagreeable.

————

April 30th.

Went to the Louvre, where we saw the Museum of Antique Sculpture. The galleries are very beautiful. The Venus of Milo, a noble statue, shown to excellent advantage, standing alone in the centre of a chamber, set off by deep red draperies. Her face has more character, and a more lofty expression, than is usual in a Venus ; but resembles in this the Venus of Capua. Admired also the Venus Genitrix, the famous Diana with the Deer ; the Diana fasten-

* Presented, I think, by the widow of Adrien de Jussieu.

1857. ing her tunic; the Gladiators or rather Warrior; a Genius of Repose; and a very noble statue of Augustus.

A few guests came in in the evening: M. de Vigny, who was very entertaining; Madame Tourgueneff; Mrs. Champion and the Byrnes.

May 1st.

Went to the Jardin des Plantes to call on M. Decaisne, but did not see him.   Looked at the fine old Cedar, planted by Bernard de Jussieu in 1755; but I think I have seen finer in England.   Then to the Louvre, saw the Gallery of Modern Sculpture, with which I was not much struck; and that of Sculpture of the Renaissance, which is interesting, Fine monumental figure of the Connetable de Montmorency.   The naked Diana with a stag, by Jean Goujon, commonly said to be a portrait of Diane de Poitiers, but this seems to be doubtful.   The face, however looks like a portrait, not at all ideal; the figure long and lanky.   A nearly similar subject in bronze, by Benvenuto Cellini.

May 2nd.

Fine day; the Horse-chesnuts beginning to flower in the Tuileries gardens.

Went to the Louvre, but not admitted, because the Grand Duke* was there.   Then to the church of St. Germain l'Auxerrois: handsome W. front, in the flamboyant style, built in the 15th century: fine rose window, and rich porch with five arches. In the interior, beautiful group of little angels

* A Russian.

supporting a cross, after a design and model by 1857.
Madame de Lamartine.—From the belfry of this
church was given the signal for the St. Bartholomew
massacre. Thence we went to St. Germain des
Prés, one of the oldest churches in Paris, but so
much restored, repainted and regilt, that it has
no appearance of antiquity, and the effect of the
interior is too gaudy. Near this was formerly the
prison of the Abbaye, notorious for the massacres of
September '92

May 3rd.

A beautiful day. Went to the Bois de Boulogne
with Fanny, and spent above two hours very
pleasantly in driving and walking about it. Great
improvements lately made in this place, which is
now a very agreeable park. Artifiicial lakes very
pretty, with well-wooded banks : rocks (masses of
Fontainebleau sandstone) brought from a distance,
and very naturally and skilfully grouped ; pleasant
woods of birch in all the fresh and tender green of
its young foliage, and of young oaks, now also
putting forth their leaves. I saw no large or old
trees. A public garden, the " Pré Catalan," in the
midst, in which we admired the most beautiful beds
of Heartsease and Anemones, the rich crimson
flowers of the Hybrid Rhododendrons, and the
white of the Magnolia conspicua. This last is
frequent in the gardens at Paris, and seems to be
perfectly hardy. In the evening we went to M.
de Lamartine's. I was glad to know him. He is a
tall, thin, handsome man, grey-haired, with a fine

1857. countenance. Madame de Lamartine, an English-woman, is a very well-informed and agreeable woman : she has a great talent for painting and modelling; her paintings on China are remarkably beautiful. Lamartine did not talk much.

Speaking of Paul Delaroche's pictures and their subjects, he remarked to me that English history is much more rich in subjects for painting and the drama than French history, and altogether, in his opinion, much more interesting. Henry the Fourth of France, he said, was a great character, but with a mixture of the comic,—much of the Gascon in him. An Englishman of the name of Alcock* came in, who had resided much in China, and in several different parts of it, both in the north and south. His conversation was entertaining.

May 4th.

Count Alfred de Vigny took us to see the Sainte Chapelle, adjoining to the Palais de Justice; it was built by Louis the Ninth, to receive a piece of the True Cross, and some other relics which he had brought from the Holy Land. The interior a very beautiful specimen of Gothic architecture, remark-able for its loftiness and the graceful lightness of the effect. It has been admirably well restored: the colouring though rich is not gaudy, and assists the effect of the architecture. Windows of prodigious height and beauty, of very fine old coloured glass. Armorial bearings of King Louis, of his wife, and of his mother, Blanche of Castile, on the walls and

* Sir Rutherford Alcock.

pillars. Little cell with a strong door, and a small 1857.
strongly-barred window looking into the chapel, in
which Louis the Eleventh used to lock himself up
to hear Mass in security. Graceful and highly
ornamental spire of the chapel. Palais de Justice :
great hall called the Salle des Pas Perdus ; monu-
ment of Malesherbes.

Recrossing the river, M. de Vigny pointed out to
us the old part of the Palais de Justice ; the grim-
looking towers with high conical roofs ; the Tour de
Nesle, with its somewhat mythical tales of terror ;
and the Conciergerie, associated with more recent
and more certain horrors. Thence to the Louvre,
where we admired the new buildings ; then to the
gardens of the Luxembourg, which are very pretty.

The lilacs not yet fully in blossom. M. de Vigny
very agreeable and entertaining. He mentioned a
curious instance of Russian barbarism, which had
happened during the levee in the Tuileries that
morning ; a Russian general, one of the suite of the
Grand Duke, violently kicking and cuffing one of his
attendants—not indeed actually in the Emperor's
presence, but in one of the state rooms ; the man
submitting without the least attempt at resistance ;
the French soldiers indignant, and only restrained
by their officers from dragging the assailant out of
the room.

May 7th.

A pleasant evening party at Mrs. Power's. Had
some talk with Madame Laugier, M. Mathieu, M.
de Verneuil, and General Ulloa. Madame Laugier

1857. (niece of the great Arago), a very agreeable woman, talked much of the recent embellishments of Paris, the enormous expenditure, the enormous increase in prices, and especially in house rent. She said the cost of living is now quite as high in Paris as in London. Much talk also about the great review, of 50,000 troops, which took place yesterday in the Champ de Mars. M. Mathieu (the father of this lady, and brother-in-law of Arago), was one of the *jurors* of the Great Exhibition in London in 1851, for the department of philosophical instruments: spoke with great admiration of the beauty of the original Crystal Palace; talked of Mr. Babbage, Sir J. Herschel, and Sir D. Brewster, whom he knew in England. Talked also of Humboldt, who was a very old and intimate friend of Arago and his family. Curious anecdote of Louis Philippe.

The last time Humboldt was at Paris, was in Dec., 1847; before leaving it he went to the Tuileries, and had a long conversation with the King, whom he had known in earlier times. As he was taking leave, Louis Philippe said to him:—" Tell my good brother the King of Prussia that I am very firm here; I am very popular; all France is at my feet; the kings of Europe may sleep soundly ('peuvent dormir sur les deux oreilles'), for there will be no more revolutions." Humboldt told this conversation to the Aragos immediately after. In little more than two months' time Louis Philippe was a fugitive.

May 8.

Went shopping on the Boulevards with Fanny;

bought two bronzes; then drove along the river-side 1857. as far as the bridge of Jena, and returned by the other side.

May 10th.

We drove to Pere La Chaise, and looked at the tomb of Abelard and Heloisa. The cemetery excessively crowded with monuments. Then to the Pantheon, or Church of St. Genevieve: a very noble building: both the outside and inside beautiful and grand. Fine inscription beneath the pediment: Aux Grands Hommes La Patrie Reconnoissante. The interior, to my thinking, the finest example, after St. Peter's, of that Italian style of architecture. In different parts of the interior are excellent copies of Raphael's Vatican frescoes.

On another side of the Place, the great public library of St. Genevieve; a stately building. Visited also the Church of St. Etienne du Mont, near the Pantheon. It is very old; the front in a singular style, a mixture (as it appears to me) of corrupt or debased Roman with Gothic. Tournefort, and several other eminent men, were buried in this church. Here also the late Archbishop of Paris was assassinated.

May 11th.

Went to the Jardin des Plantes, saw M. Adolphe Brongniart again, and had some good talk with him. Talked of the botany of the Cape, of Egypt, of Abyssinia, and of Madagascar; of Hooker's "Flora Indica," of which he lamented the interruption; of

1857. the difficulty of cultivating certain families
of plants, and the plants of certain countries;
of the great proportion of Conifers in cultivation, M. Brongniart said, that the most
difficult plants to cultivate are those of elevated
regions in hot countries, such as those from
the *paramos* of the Andes, and from the *campos* of
Brazil. He remarked that we have no complete
flora of any tropical island, not even of any one of
the West Indian islands, nor of Mauritius; and that
such a flora is a great desideratum in geographical
botany. I then visited, with a ticket given me by
M. Brongniart, some of the hot-houses, and the
orangerie or cool greenhouse: this last in particular
very inferior to those at Kew. Then to Bailliere's,
the natural history bookseller, near the Ecole de
Medicine, and bought two of St. Hilaire's works on
Brazilian botany.

_____

May 12th.

A beautiful day. To Versailles. A very pleasant
drive of an hour and a half, through the Bois de
Boulogne, crossing the Seine, ascending the steep
heights of St. Cloud, from which we had an admirable view over Paris and the extensive plain; then
through pleasant woods most of the way. The
approach to Versailles striking, but the Palace does
not show well from the Place d'Arnes; one does not
at first at all appreciate its magnitude. After
luncheon at the Hotel of France, we went into the
palace; saw the chapel and the theatre, and an endless succession of rooms and galleries, with acres of

battle-paintings; not a little fatiguing. Horace 1857.
Vernet's battles are, however, finely painted, especially the capture of Abd-el-Kader's camp ("smalah")
which is a very striking composition; but I had not
time to study it sufficiently. The grand gallery, or
Galerie des Glaces, is most magnificent. We were
more interested by seeing the private apartments,
in which Louis the Fourteenth and his successors
lived, and especially the suite of little rooms—
curiously small—inhabited by poor Marie Antoinette,
and in which she was so nearly murdered by the mob
in October, 1789. The gardens of Versailles are
certainly very fine, in their way perfect types of that
style; the aspect of the palace as seen from them
is very imposing; and, altogether, the palace and
gardens both have exactly that sort of grandeur
which appears appropriate to their history—characteristic of the Grand Monarque and his time. The
numerous statues, many of them copied from the
finest antique works, have a very agreeable effect
amidst the alleys and hedges of clipped hornbeam.
Thence through the Park of Versailles to Le Petit
Trianon; were too late to see the palace, but strolled
in the gardens, which are delightful, and very
different from those of Versailles. Bernard de
Jussieu is said to have been employed to lay them
out. They are entirely in the English style; like a
fine specimen of an English gentleman's grounds;
abundance of shade and verdure, noble trees grouped
or scattered as if by nature; mossy turf, flowers,
wood, and water, without formality. Many of the
trees are remarkably fine; in particular, the largest

1857.   Sophora Japonica, the finest Weymouth Pines, and
some of the finest Planes, I ever saw. This is the
place where poor Marie Antoinette spent the gayest
days of her life.

_____

May 14th.

Went to the Ecole des Mines ; spent an hour and
a half there, examining the collection of minerals
and fossils, which is very interesting and instructive.

Afterwards I walked in the gardens of the Luxembourg, which are now in great beauty ; the lilacs
in profuse blossom. The Judas-tree, which is very
common in the gardens of Paris and its neighbourhood, and grows to a large size, is now perfectly
covered with blossoms, extremely beautiful.

In the evening we took a drive in an open carriage
round a great part of the Boulevards, and through
the Place de la Concorde, over the bridge of that
name, and along a part of the quays,—delighted
with the beautiful effect of the innumerable lights,
and with the gay and brilliant scene of the Boulevards, with their shops, cafés, and theatres.

_____

May 15.

The weather still delightful. We went by railway
to St. Germain en Laye, and spent some hours
there very agreeably. The situation is fine : the
palace and town standing on a plateau which rises
abruptly above the Seine. The noble terrace running for a great way along the brow of these heights
(a work of Le Notre, completed in 1696), commands

an extensive and very agreeable view over the plain 1856.
in which Paris is situated : the city itself is not very
distinctly seen, but the heights of Montmartre are
conspicuous; nearer, the bold eminence of Mont
Valerien, crowned with a large fort, has a fine effect
in the view; and more to the south, the range of
hills towards Marly.

The windings of the Seine, below us, are beautiful.

The old chateau, or palace, in which our James
the Second spent the last years of his life, is a large,
gloomy, melancholy building, now used as a prison ;
but the situation is far better than that of Versailles.
The chapel of the old palace, in which Louis XIV.
was baptized, now forms part of the *restaurant* called
the Pavillon de Henri IV.; it is situated on the edge
of the terrace, and commands a delightful view.
Adjoining to the chateau and the terrace is the
public garden, laid out not long ago, and very
agreeable : beautiful flowers and large shady trees.
We spent two hours and a half very agreeably in
driving in the forest; the fresh tender green of the
young foliage and the refreshing shade in this brilliant weather, were exceedingly pleasant. The forest
is composed for the most part of young and slender,
though pretty tall, trees (being cut down at regular
intervals) ; oak, beech, birch, and a great deal of
hornbeam ; here and there some oaks of a good age
and size. We were shown two very large oaks, fine
and venerable trees : one the Chêne de St. Fiacre,
as large as almost any oak I have seen ; the other,
la Chêne de Notre Dame du Bon Secours, somewhat
less, but a fine tree. The only wild plant, not com-

1857. mon in England, that I saw in the forest, was the
Solomons' Seal, Polygonatum multiflorum.   Blue-
bells in great profusion.

_____

May 16th.

Weather magnificent.

Visited the Louvre, and spent some time in the
gallery of antique sculpture, admiring especially the
Venus of Milo, the Venus of Arles, the Venus
Genitrix, the Diana fastening her tunic, and the
Jason putting on his sandal.

_____

May 17th.

We visited M. de Verneuil, had much pleasant
talk with him, and saw some of his fossils, principally coal-plants collected in Spain. He has travelled
extensively and repeatedly in that country, showed
us some interesting tables of barometrical and ther-
mometrical observations which he had made there;
showed us the mountain barometer which he had
used. He mentioned that in one place in Andalusia,
he had seen the Agave Americana growing at an
elevation of 900 metres.   In the garden of his
house I observed a fine tree of the Flowering Ash,
Fraxinus Ornus, in full blossom. The temperature
of this day, at the time we were at his house, was
23 deg. centigr.

We went to the Louvre, and looked through the
gallery of sculpture of the Renaissance ; were much
struck with the works of Jean Goujon, and some of

his contemporaries ; particularly a "Deposition from 1857.
the Cross," and "The Four Evangelists," by Goujon.
The monumental figures of the Connetable de Montmorency, and of Philippe de Chabot, Admiral of
France, are fine. Two figures of captives, attributed
to Michael Angelo, have a certain grandeur. We
went in the evening to M. de Lamartine's.

May 18th.

M. de Verneuil took me to a séance of the Acadèmie
des Sciences.

Professor Matteucci of Pisa was elected a corresponding member. The morning was very fine
and hot, a storm came on about 4 p.m., but did not
last long. A very pleasant evening party at Mrs.
Power's : — Madame Tourqueneff, M. de Vigny,
General Ulloa, Professor Sismonda of Turin (whom
I had visited in the morning), Mr. and Mrs. Byrne,
M. de Vigny singularly agreeable : he has most
attractive manners, and the precision and beauty, as
well as fluency of his language, are remarkable. I
never heard anything better than his talk this
evening upon dramatic matters, particularly upon
Madame Ristori's acting, upon the character of
Alfieri's Myrrha, upon Alfieri's dramatic genius ; the
difference between the character of Myrrha in Ovid
and Alfieri's treatment of it, and the corresponding
difference between the Phædra of antiquity and
of Racine.

May 21st.

Weather delightful. Drove out with Fanny, and

1857. after some shopping, we went to see the Tower of St. Jacques de la Boucherie, situated near the Place du Chatelet. The improvements in that part of Paris have cleared away the mean streets which encumbered it, and laid it open most advantageously to view; a pretty little public garden has been formed round its base.

It is a very tall and graceful tower, a remarkably beautiful specimen of Gothic architecture: it was the tower of a church, of which it is now the only remains. The ground story is open, and under its arch has been placed a fine statue of Pascal. Thence we went on to look at the exterior of the Hotel de Ville, which is now nearly completed, and is a noble building. The original building is one of the most striking specimens of the Renaissance style, and the additions are excellently well in keeping with it. The statues of eminent Frenchmen, which form a series along the west face, are very good.

May 25th.

A succession of violent storms of rain and thunder throughout the afternoon.

We went to the Theatre Français, saw a lively comedy by Scribe:—"La Bataille de Dames;" excellent acting by Prevost, and Madlle. Nathalie. M. de Vigny joined us in our box, and afterwards accompanied us home, and was very agreeable. Story of the performance of this play before the Court at Fontainbleau, the other day: the Emperor had the performers presented to him, and said to

Prevost (who acted the part of a Prefêt, M. de Montrichard) :—"M. Prevost, you make an admirable Prefêt." — "I am quite at your Majesty's service, if there should be occasion for me."

"No, no, M. Prevost, better remain as you are; it is much easier to find a good Prefêt, than to find such an artist as you."

M. de Vigny told us, that General Canrobert, with whom he is acquainted, always speaks highly of the English generals, as well as of the English army, in the Crimea, and holds them (the generals) to have been very unjustly and unreasonably attacked; that he considers them to have acted under peculiar disadvantages and difficulties owing to the interference and control of the newspapers.

————

Visited the Bibliotheque Imperiale, and looked at the beautiful collection of engraved gems.

A visit from M. de Vigny in the evening: he made Fanny a present of his poetry, in a neat volume, and read aloud some of his poems to us, admirably well; I never before was so much pleased with French poetry, nor could have believed that it could have so agreeable an effect on the ear. His anecdote of a great rich Jew speculator (one of those three or four Jews who are at present the great millionaires of Paris) :—this man remarking that the conduct of *Cain*, as shown in the Bible, was very bad ;—the person with whom he was talking, appearing to think this somewat of a truism, the millionaire added :—
"I mean that he was a very bad speculator."

1857.   "A bad speculator! I own I never looked upon
him in that light."

"Yes: when he found that Abel was in favour with
the higher powers, he ought to have turned that
fact to account, to have profited by his interest, to
have *utilisé* him, instead of knocking out his brains."

M. de Vigny's account of the Jesuits, particularly
their system of education; the training of a Jesuit,
calculated to develope and cultivate the particular
talents of each one, but at the same time, and above
all, to keep the mind in the most absolute subjection,
to subdue every trace of independence of character,
and to produce, above all things, the principle and
habit of the most absolute obedience. Thus, for
instance, if a Jesuit had a particular turn for reading
and study, his superiors would sometimes command
him to abstain for a certain time from looking at a
book, or would send him to minister to a particularly
illiterate population; if he became famous as a
preacher, they would silence him for a time; and all
this to cultivate the habit of implicit obedience and
entire abnegation of will.

_____

May 30th.

The 13th anniversary of our happy wedding-day.
The weather raw, blustering, and chilly. Spent
much of the day in book-hunting in the quarter of
the Ecole de Médicine; bought Humboldt's "De
Distributione Geographicâ Plantarum."

_____

[In the beginning of June the Charles Bunburys 1857. left Paris and returned to England, accompanied by Mrs. Power as far as Folkestone, when she proceeded to London, and they went to Sandgate to pay a visit to Colonel and Mrs. Bunbury and their two children, where Col. Bunbury had lately obtained a Staff appointment].

LETTERS.

Sandgate,
Sunday, June 7th, 1857.

My Dear Lyell,

I saw Adolphe Brongniart several times while we were at Paris, and had a good deal of conversation with him ; and I had the advantage of seeing the magnificent botanical galleries of the Jardin des Plantes under his guidance. But he seems pretty cool about paleo-botany, and does not hold out any promise of further publication on the subject. He is still sceptical about the *Antholites* of the coal formation, though without having seen either Lindley's or Prestwich's specimens, and of course knowing nothing of mine except from what you and Hooker have said of it in the supplement, and what I told him of it, he says he shall wait with much curiosity for a full description and figure of it from Hooker, in whose accuracy he has especial confidence. He suspects that the Antholithes Pitcairnicæ may be the spike of fructification of some Asterophyllites or Annularia (probably be-

1857. longing to the class of Gymnosperms) ;—that the supposed flowers may be clusters of bracts, enveloping a seed or nut. There is undoubtedly some plausibility in this theory, and as far as I remember it is particularly applicable to Prestwich's Antholites; I should like to know what Hooker thinks of it. Brongniart said that he purposely omitted Antholites in his Tableau des Genres, not thinking the evidence sufficient. To go to quite another subject : M. Brongniart entirely disbelieves the origin of Wheat from the wild Ægilops of the South of Europe : all, he says, that he can admit as the result of the experiments is, that Ægilops and Triticum may probably be the same *genus*, and that hybrids may be formed in cultivation.

I saw also at Paris, M. Barrande and M. de Verneuil, both of whom were very civil to me ; also M. Sismonda of Turin. They all talked of the grand puzzle of the Alpine Anthrocite schists, on which it seems there had lately been a great deal of discussion in the French Geological Society, and on which my researches had been quoted. They, at least, M. Barrande and M. Sismonda, seem inclined to attach more weight to the Jurrassic character of the animals than to the carboniferous character of the plants ; and in truth, if it be certain that they really alternate, and that there is no folding or displacement (and all the foreign geologists who have visited the locality seem very positive as to this)—then they are probably right in trusting more to shells than to plants. At any rate, it remains, as they all admit, a most puzzling anomaly and still

a solitary case. We were both much pleased with
M. Barrande : he is a very agreeable man and
evidently one of great ability. He showed us many
of the beautiful plates prepared for his great work,
and explained the characters of the fossils in a
remarkably clear and agreeable way ; particularly
pointing out the curious and interesting series of
changes which he had traced in Trilobites, and also
in the shells of some of the Cephalopoda, from the
earliest stage to maturity. He showed us also
reports that he received regularly from his Bohemian
workmen. He spoke with great pleasure of your
visit to him.

In the Museum of the Ecole des Mines, I admired
the fine collection of those strange shells, the Hip-
purites (the finest set of them, M. de Verneuil told
me, that is to be seen anywhere), and the magnificent Ammonites in red marble, from the Hallstadt
beds of Austria, which interested me particularly
from recollecting the accounts of them in your
letters last autumn. M. de Verneuil showed me a
good many fossil plants (carboniferous) which he had
collected in Spain, but there was nothing new to me
among them ; the only point at all remarkable was
the abundance of a Dictyopteris, a fern with reticu-
lated veins, almost exactly like one found in North
America.

My trip to Paris has not been unprofitable in the
way of science, though more of my time, perhaps,
was given to other things. We certainly enjoyed
our stay there exceedingly, and came away with
most agreeable impressions of the great city ; I do

1857. not wonder the French are so proud of it.—How
are you getting on with Madeira and Teneriffe?
Is the *magnum opus* to make its appearance this
summer? I should like to hear something of what
you have been about all this while, and whether
there has been anything interesting at the Geological
Society. I hope Mary is very well; pray give my
love to her, and to the rest of the family, especially
the Pertzes; and Katharine, to whom I owe a
letter.

Ever yours affectionately,

C. J. F. BUNBURY.

Sandgate,

June 9th, 1857.

My dear Katharine,

I thank you much for your kind letter of
the 4th. I am very sorry we cannot be with you
to-morrow,* I should like it very much, and I feel
with you as to the comfort of those occasional family
gatherings; however, my best wishes and hearty
affection will be with you. I daresay you are very
glad to have Mrs. Power with you, and that it is an
equal pleasure to her to be among you all. She is
the kindest and most unselfish of human beings. I
can never say enough of her goodness to us all the
time we were at Paris. We spent a delightful time
there which I shall always remember with pleasure.
There is a wonderful variety of beauty and interest
in that city. I do not wonder at the admiration

* Mr. and Mrs. Horner's wedding-day.

with which the French always speak of it, nor that 1857.
they should exalt it above all other cities.

We shall probably remain here a fortnight ; I am
afraid we shall not be able to stay long in London,
for various reasons, but I hope we shall have a long
visit from you and all your party at Mildenhall.

M. Brongniart, at the Jardin des Plantes, showed
me what was very interesting to me, a part of
Tournefort's original herbarium, which is still kept
separate, and also of the herbarium of his contem-
porary Vaillant, which formed the basis of their
great general collection.

Tournefort's specimens, collected more than 150
years ago, are still in good preservation, though
many of them very imperfect; Vaillant's nearly
equally old are much finer, and in excellent condition. There is something very interesting to a
botanist in seeing these ancient *typical* collections—
the very specimens which have been seen and
handled by the early masters of our science.

With much love to all the family party, I am ever
Your affectionate brother,
C. J. F. BUNBURY.

***

Mildenhall,
Saturday, July, 1857.

My dear Mr. Horner,

Lyell's letter on the glaciers is indeed a
very remarkable and important one—quite a treatise,
fit to make a chapter in the Principles. I have let
Babbage read it, as he is staying with us, and he is
very much struck with it. It is curious that Babbage,

1857. before reading it, on my telling him that Lyell seemed now disposed to adopt the theory of a gigantic glacier extending across the great valley of Switzerland, suggested that the Alps may have lost in height since then—the very point that Lyell urges in his letter.

I am very glad the Lyells are going to Italy; it will be a great pleasure to Mary, though the time they allow themselves is, perhaps, too short for her to have the full enjoyment of it; at least, it is much too short according to *my* notions of travelling; but there are different notions on that subject. Still she will see much that will delight her; and I have no doubt that Lyell will gain valuable hints from a renewed visit to Vesuvius and Etna. I almost envy them their journey.

We have had a very pleasant visit from Babbage and the Bowyers: the former in a very agreeable mood, very conversible and amusing, and with less than his usual misanthropy; very cheerful and friendly. He and Mr. Bowyer seemed to take much to each other. My niece and nephew, Louisa and Harry, are also with us now. Fanny pretty well, and, as usual, excessively busy. My father, I am sorry to say, not materially better. I hope Folkestone continues to agree with Mrs. Horner and all of you.

Of the horrible Indian business what can one say but that we feel deeply thankful to have no personal friends in that country. The details in the newspapers sicken one with horror. What must they be to those whose friends were among the victims? It

is very evident that a principal cause—if not of the 1857.
revolt itself, at least of the peculiarly atrocious
character it has assumed—has been religious hatred
and bigotry of the Mahommedans, which so often
gives a licence, and an apparent sanction, to the
worst passions of mankind. After the country has
been reconquered (for of our ultimate success I do
not doubt), and exemplary punishment inflicted on
those monsters at Delhi, it will then be a great and
most serious question how such a country is to be
governed for the future.

With much love to Mrs. Horner, and Susan and
Joanna, I am ever

Your affectionate Son-in-law,

C. J. F. BUNBURY.

Mildenhall,
August 22nd, 1857.

My Dear Mr. Horner,

I am much obliged to you for sending me
Lyell's two letters, which are indeed full of scientific
interest, and of important and curious matter. It
is very important that he has satisfied himself by
actual examination, of the true position of those
Aix-la-Chapelle beds, so rich in fossil land plants.
Without such evidence one should have had a
difficulty in believing that such a deposit was older
than the tertiary age. It is a very remarkable and
important fact. I saw Dr. Debey's collection, at
least a good part of it in '55, and a wonderfully
rich one it is; he told me then that he intended to
publish a description of it, and this I hope will soon

1857. be done. But I confess I did not feel so well satisfied
of the certainty of all his determinations, as Lyell
appears to be ; but he did not appear well disposed
to listen to any objections or cautions. With
respect particularly to that remarkable family of the
Proteaceæ, of which Lyell speaks so much—some of
the leaves that Debey referred with confidence
to it seemed to me very doubtful. Still there
can be no doubt that these remains prove the
existence of a very rich and varied vegetation, quite
unlike that of Europe in the present day, equally or
still more unlike that found fossil in the Jurassic
rocks and consisting in great part of that class of
plants (the dicotyledonous) which till lately was
supposed not to have come into existence before the
tertiary period.

Heer of Zurich must be a remarkably clever man
from the great influence he seems invariably to gain
over all who come in contact with him. I see he has
made Lyell completely his disciple and a believer in
his interpretations of fossil leaves. I shall be very
curious to hear what Joseph Hooker says to this
matter, for I certainly consider him a higher authority
on botany in general than any one now living on the
Continent. I have in a Memoir I have just finished
for the Geological Society (a notice of Lyell's
Madeira Plants) entered somewhat into this question
and given my reasons for doubting whether plants
can be determined with so much certainty from the
leaves alone, as the Swiss and German geologists
contend ; but after having read these letters, I shall
carefully revise what I have written. I was not

aware before that many fossil leaves had been found 1857. in a sufficiently perfect state for the *stomata* (pores of the epidermis) to be examined.

I have been for some time past making a careful examination of the leaves of various recent families, in reference to their veining, and I am now turning my attention to the stomata also. But I am a slow worker. How active and indefatigable Lyell is! What zeal and energy! he is really a model man of science. The picture that Kingsley draws (in "Glaucus") of an ideal naturalist, is very like him.

I am most truly glad to hear such continued good accounts of Mrs. Horner; we hear so much of her increased strength and of the good that Folkestone is doing her, that it seems not unreasonable to hope she may recover entirely before the winter comes on. I hope Susan also is gaining strength—I am sure she wanted it.

The gault and green sand of Folkestone and its neighbourhood will probably supply you with some occupation and amusement: Joanna, I am afraid will find very little in the way of "wonders of the deep," for I have seldom seen a shore that appeared more barren.

My Father's illness has been very long and distressing and his recovery very tedious, nor can I say that he is yet by any means well, but I trust that he *is* recovering, though slowly. Mrs. Power's company has been a great pleasure to us, and I hope we have been able to make the time pass pleasantly for her, though we have had but little society, and the

1857. country hereabouts does not furnish many resources. The Pellews and D'Oyleys have just left us.

With much love to Mrs. Horner, Susan and Joanna, I am ever ·

Your affectionate Son-in-law,

C. J. F. BUNBURY.

Mildenhall,
September 1st, 1857.

My dear Emily,

I am exceedingly sorry to hear such an uncomfortable account of my dear Father, and that he is again suffering so much from his most obstinate and troublesome enemy. I hope that this renewed attack will pass off in a few days; but it does indeed make one very anxious, the continuance of such a harassing complaint at my father's time of life.

The accounts from India are indeed horrible enough, and show more clearly what ferocious enemies we have to deal with. The character of religious hatred and bigotry is becoming more and more evidently stamped upon the rebellion, and it is plain that our enemies intend to make it a war of absolute extermination. General Havelock's successes and Brigadier Nicholson's, and the uniform ill success of the rebels in their sorties from Delhi, are comforting; and, indeed, Cawnpore seems to be the only success that the wretches have gained, otherwise than by mere surprise. It is a fine thing, our troops marching and fighting so well in such dreadful heat.

I am sure those who survive the campaign ought 1857. to have all the honours and rewards that a nation can bestow. I trust the next mail will bring an account of the relief of Lucknow and the punishment of Nena-Sahib, if not of Delhi.

I am glad you were pleased with Hermann Pertz; he is a great favourite of ours; a fine young man, I think, intelligent, gentle, and spirited. He is the youngest of Chevalier Pertz's sons.

Believe me,

Your affectionate stepson,

C. J. F. BUNBURY.

Mildenhall,

October 4th, 1857.

My dear Joanna,

Many thanks for your very agreeable letter of the 1st, and especially for your interesting account of the Manchester Exhibition of Pictures, which I am very glad you have had the opportunity of seeing. I have no doubt, from all I have heard, that it is a really noble and very interesting collection, and I wish very much that Fanny and I could have seen it; but the fates forbid; for many reasons it is inconvenient to us to leave home this month, so I must be content with hearing of it; and it is a great pleasure to have such a good account as you have written me of it. We have had one of the gayest and merriest and pleasantest parties in our house for the last week that I ever remember; nothing but laughing, nonsense, and jollity from

1857. morning till night, and till a pretty late hour of the
night too. I only wish you or Susan could have
been of the party. I do not know whether *you* know
the MacMurdos. I know Susan does; I have got
much more thoroughly acquainted with them this
time than I ever was before, and I am delighted
with them. *He* is a *very* clever man, very pleasing
in his manners, and a true gentleman; and *she* is
worthy to be Sir Charles Napier's daughter. Their
daughter Kate is a very nice little girl of ten years
old—the only one of all their ten children that they
have brought with them. Our party is now breaking
up; the MacMurdos went away yesterday; the rest
are going to-morrow, and to-morrow evening we
shall be all alone again, which will seem quite
strange to us. Fanny Pellew has been here all this
week, and seems to have enjoyed the party exceedingly; she is really a very charming girl.

(*Tuesday, October 6th.*) My letter was interrupted
by my having to go suddenly to Barton on Sunday;
my Father had had an attack of palpitation of the
heart which alarmed Lady Bunbury, but on my
arrival I found him quite recovered, and he was in
very good spirits that evening, talking with the
MacMurdos and enjoying their conversation. I
think he is certainly better in health on the whole,
but still very lame, and very weak on his legs.
There was important business to be transacted at
Barton yesterday afternoon, so I did not get back
till late in the evening, and then found Fanny quite
alone. It feels almost strange to be alone again
after ten days of such continual racket and gaiety:

but we have plenty to do, and are not likely to be tired of each other, but on the contrary intend to be very happy in our quiet seclusion for some time to come. Our glorious summer is over, but we have had lately many beautiful days—fine autumn days, calm, bright, and sunny, though chilly in the mornings and evenings, or in the shade. In this sort of weather, I think the country appears more beautiful than even in the finest summer; there is something peculiarly soft and delicate in the colouring, especially of the distance.

This fine and hot summer has caused many garden plants here and at Barton to bear fruit for the first time : such as the Catalpa, the Bignonia radicans, and others. The crop of acorns has been enormous. But I have seen no remarkable abundance of butterflies or other insects this year, nor any uncommon kinds.—I am sorry, but not at all surprised that you found the shore at Folkestone so unproductive of shells and other marine curiosities : it looked to me exceedingly unproductive ; an open shingle beach without rocks is almost always barren. You would be delighted if ever you went either to Scarborough or to Ventnor with their marine animals and plants. But I am very glad that Folkestone answered so well as it seems to have done for the main object of Mrs. Horner's health. The accounts of her are most cheering, and I rejoice most sincerely in them.

I do not know whether you have heard lately from Edward. I had an entertaining letter from him when we were at Norwich, written from a place

1857. called (I think) Darness, in the extreme north of Sutherland, and giving an account of Cape Wrath and that wild country. My father has since had a letter from him, written after his return to George Lock's from the wild north-west country, with which he had been much pleased. By the bye, I forgot you may probably see him before I shall, as he meant to revisit Manchester on his way back to London.

We spent three days very pleasantly at Norwich, with Sedgwick. He was extremely kind to us, and I never knew him more agreeable. Besides the cathedral and the Roman camp at Caistor, we visited with him three chalk-pits where we saw fine examples of the curious sand-pipes described by Lyell in his Manual: the enormous vertical fluits called *pot-stones*: the overlying stratified sand and gravel, and in one spot the *crag* full of shells resting immediately on the chalk. You have been at Norwich, I know, and doubtless saw all this. I was was much pleased to see it.

I have this morning been reading Lyell's very interesting and very learned letter from Genoa.

Pray thank Mr. Horner for sending it to me : I will return it to-morrow.

The accounts from India continue to be most painful. I wish the report from Calcutta which I read in the *Evening Mail* this morning may prove true,—that the rebel forces before Lucknow had broken up, and retired ; but I much fear we shall hear of other horrible disasters before our army can be assembled. It is altogether a dreadful business. I should still feel confident of our reconquering

India if it depended purely upon military operations, 1857.
but I am afraid the civilians at Calcutta will spoil
the whole. To-morrow is to be a "day of
humiliation." If Lord Dalhousie and the Court of
Directors, and the supreme Council of Calcutta
would fast and humble themselves in any amount of
sackcloth and ashes, it would be very appropriate.

I have been reading an exceedingly able and
vigorous pamphlet on the subject which I strongly
recommend to you if you wish to know the origin of
all this : it is entitled " The Mutiny of the Bengal
Army," an Historical Narrative by one who has
served under Sir Charles Napier."

With much love to your Father and Mother, I am
your very affectionate Brother,

C. J. F. BUNBURY.

JOURNAL.

October 7th.

I remarked one day in conversation about the
Indian affairs, how advantageously the character of
Akbar Khan (the Cabul chief) came out by the
contrast with these Indian ruffians who have per-
petrated such unspeakable atrocities at Delhi,
Cawnpore, and other places. Susan MacMurdo
coincided with me, and Colonel Hardinge said that
the Afghans are indeed a fine chivalrous people,
very superior to the Hindoos.

1857.   October 8th.

In the three months and a little more that we have been at home, since our return from London, we have hardly been a week alone. We have had several very pleasant and some interesting guests.

LETTERS.

Mildenhall,
October 15th, 1857.

My Dear Katharine,

I do not think I have written to you, nor heard *directly* from you since you went to Scotland. It was a real relief and comfort to hear that you had determined to return by land and not by sea ; for the weather was so very stormy and bad just at that time, that it made us very uncomfortable till we heard of the change of your plans.   Ever since, we have had very still weather, sometimes rainy, but with little wind ; as if Æolus, disappointed of catching you had not thought it worth while to make any more bluster.   I was very glad to hear of your safe arrival at home, and have no doubt that you are very happy to be there again, and find plenty to do.   I hope you enjoyed your stay in Scotland. Did you do anything in the way of botany there ? I suppose you made no discoveries—no more Buxbaumias or the like—or I should have heard of them.   I hope Leonard cultivated his botany a

little. I cannot myself boast of any discoveries, but 1857.
I have not been entirely idle.   I have finished a
catalogue of my Ferns ; have revised repeatedly,
and with great care, my paper on the fossil plants
from Madeira, and added so much to it that it is
equivalent to a new paper ; have written another
Memoir for the Geological Society on a genus of
fossil Ferns ; made some progress in cataloguing
the plants in my possession from Buenos Ayres and
Uruguay, with a view to a paper at some future
time ; continued my studies of the characters of
leaves ; read the third and fourth volumes of Sir
Charles Napier's Life ; and am now reading that of
Sir Thomas Munro.   I have moreover, chopped not
a few trees.   Here is an egotistical chapter !   We
spent three days very pleasantly with old Sedgwick
at Norwich, where he lives quite as a *family man*—
bachelor though he be—with his married nephew
and his pretty little wife and their children, and an
unmarried niece besides.   Since then we have had
an exceedingly agreeable, joyous, merry, madcap
party : our house constantly full ; the MacMurdos,
and William and George Napier permanently ;
Fanny Pellew almost the whole time ; Mr. and Mrs.
Pellew, General Simpson, Colonel Harding, Mr.
and Miss Sedgwick, as birds of passage.   I never
remember a merrier or more good humoured party.
I am delighted with the MacMurdos, with whom I
greatly improved my acquaintance ; and William
Napier is quite as a brother to us both.
    We have had a glorious summer, almost the
finest I remember, and the early part of the Autumn

1857. very fine too. Fanny and I have greatly enjoyed it.
My Father has unfortunately been, as you already
have heard, a great invalid and a great sufferer the
whole summer, and is still far from well, but I hope
is really recovering, though very slowly. I have had
to go over to Barton almost every week to see him.
Henry has had a very serious illness, but is now
well again. Edward just returned to town. We
both miss the Evanses very much. What a
delightful tour Charles and Mary are making! I
really can hardly help envying them, and when I
read of their being at Genoa and Pisa, and above
all Florence (the most charming town in the world,
to my thinking), it makes me quite restless. Mary
seems to be enjoying it thoroughly, though they
travel faster than would suit me.

I do not know anything so delightful as travelling
in Italy with plenty of money and plenty of leisure.
I dare say they are just now enjoying a beautiful
clear blue sky, instead of the dull, heavy, lowering
weather that we have here.

The last news from India seems on the whole the
best that we have yet had. Lucknow re-provisioned
—our troops before Delhi healthy—the defeat of
the body of rebels that attempted to intercept
the siege train,—and no fresh disasters reported
anywhere, are comforting. If the scattered handfuls
of our brave fellows can hold their ground till the
forces from England can take the field, it is all over
with the rebels. The only consolation in the whole
of this dreadful affair is the heroism that our
countrymen have shown in the most desperate

situations. It is altogether a horrid business, and 1857. not less so because, however the struggle may end, it cannot fail to leave a feeling of deep and lasting hatred. Generations must pass away before the English can forget the horrors of Delhi and Cawnpore, nor probably will the Brahmins and Mohammedans soon forget their temporary triumph. And when all ostensible resistance is put down, I fear the country will long be insecure, long infested by bands of savage robbers. If it were purely a military question, I should feel little doubt of our ultimate success, but I am afraid of the civilians at Calcutta. If I could have my own way, I would suspend (I do not mean *hang*), all the civil authorities and make the Commander-in-chief Dictator for a year, with the absolute power of a Roman Dictator.—

" In seasons of great peril,
'Tis good that *one* bear sway."

You must have felt particularly all the horrors of this time, from your knowledge of the country and the people, and I suppose personal knowledge of some of the unfortunate sufferers. To all who had relations or dear friends in India, all this time must have been beyond expression dreadful ; bad enough to all who can love or feel for their country. And supposing we re-establish our authority, what a task it will be for the wisdom of the nation to construct a better government for British India.

Much love to Susan, Harry, and your children.

Ever your very affectionate Brother,

C. J. F. BUNBURY.

                                          Mildenhall,
                                       October 17th, 1857.

My Dear Lyell,
        I have got the new numbers of the
Geological and Linnean Societies' Journals, and
have read Falconer's elaborate and improved paper
on Elephants and Mastodons.   He seems, as far as
I can judge of the value of the evidence to make
out very clearly the specific distinctness of the
Crag Mastodon from Mastodon angustidens, and
is a true Pliocene animal.   Bell's anniversary
address to the Linnean Society is very good.
        Yesterday we had the pleasure of reading Mary's
very agreeable letter from Catania, which your
sister sent us to read ; it entertained me particularly
and though I have not been in that part of Sicily, I
recognize the truth of almost every particular in her
description, as applying equally to those parts which
I *have* seen ;   the enormus Cactuses (the largest I
ever saw) the huge horns of the oxen, the half-wild
black swine, the excessive dirt of the people, &c. ;
only the painted carts I do not remember.   I rather
suppose, from what Phillipi says of Etna, that the
dwarf fan Palm, Chamœrops humilis, which is so
very common all through the South of Sicily, does
not grow much on the east side ; I think Mary
would have mentioned it if it had been abundant.
I hope you made out your tour of Etna safely and
satisfactorily, and found much to interest you.
        Since I last wrote, there has been very important
news from India: the capture of Delhi and the relief of
Lucknow, both splendid exploits, and of the highest

importance, but purchased, I fear with many 1857. valuable lives. Two losses that we already know of are very serious—General Nicholson at Delhi, and General Neill at Lucknow. I am not so sanguine as many are, who think that the war is virtually terminated by these successes. The capture of Delhi, and of the old king, will indeed be great discouragements to the rebels; but most of them seemed to have escaped from the city, and the numbers still in arms are so great, that I do not see why they should not keep up a desultory warfare for a long time; a dreadful look-out for the poor peasantry of the country. However before this we shall have plenty of troops in India.

(This is unfinished, the other sheet wanting.)

————

Mildenhall,
October 22nd, 1857.

My Dear Mr. Horner,

I hope you will by this evening be safely settled in your comfortable home, and that Mrs. Horner will not suffer from this damp and chilly weather.

I am much interested by the accounts you give of your Egyptian researches, and very glad that they are in such a state of forwardness. The great mass of facts of a very novel kind, which you have collected and arranged, will of itself be a most important addition to our stock of knowledge; and I feel satisfied that whatever deductions you may see reason to draw from them will be most carefully

1857. considered, and worthy of the utmost attention.
I shall be delighted to see the result of your
labours. I have not read anything of Cicero's letters
since I was at college, and then but a small portion
of them : but I believe they are very interesting.
Middleton has made great use of them in his Life
of Cicero, which I read with great delight while I
was detained at Edinburgh by Fanny's illness. I
remember that one of the things which most struck
me in reading that book as coming out in the
strongest light from all the facts related, was the
excessive corruption and villainy of the *judicial* body
at Rome, in the latter days of the Republic ; and
there can hardly I think be a worse vice in the
internal state of a country. That, at least, is an
evil from which we in England have for a long time
been very free.

I quite agree with you in liking and admiring
Arnold's Roman History. I think he is in point of
style one of the very best of our modern writers, the
moral tone of his work is delightful. It was not
his fault that there is much that is heavy in the 1st
and 2nd volumes : it is inevitably tedious work to
grope for the scattered grains of historical truth
amidst much accumulations of romance and error
and confusion : conjectural history has neither the
charm of romance nor that of exact knowledge : but
when he comes to the war with Pyrrhus, to the
Punic wars, and above all to the second Punic war,
Arnold makes us full amends; his third volume is
one of the most interesting historical narratives I
have ever read. It is a great loss to the world that

he did not live to complete his work. Some find 1857. fault with him for following Niebuhr too implicitly in the early history of Rome; I do not know what degree of justice there may be in this criticism.

I am at present reading the Life of Sir Thomas Munro. Several years ago I read more about India than perhaps usually enters into the studies of those who are not in any way connected with the country; and now I have taken up the subject again, wishing to fill up the gaps in my knowledge. The book I am engaged upon is in great part very dry ,long details about the revenue of particular districts: but I find also much that is worth remembering. Munro was a very remarkable and eminent man; with wonderful powers of work, great sagacity, great determination and very high principles. He was one of the men fitted to gain and to govern an empire; we need not look far now-a-days for men qualified to lose an empire.

I hope we may be able to take advantage of Charles and Mary's kindness about their house: but we expect some friends here about the middle of November, and cannot conveniently leave home before the end of that month. Fanny is so-so. The gloom and excessive damp of the weather we have had lately are depressing, but we must not complain after such a splendid summer.

With much love to Mrs. Horner and the sisterhood, I am ever,

Your affectionate Son-in-law,
C. J. F. BUNBURY.

1857.

JOURNAL.

October 24th—26th.

At Barton where we met the Richard Napiers.

My Father mentioned in talking of the *Rolliad*, that he had formerly heard old Sir Robert Adair mention (I think it was at Mr. Ridley Colborne's)* *who* were the principal authors of that capital political satire. The chief writer was Lord John Townshend; Sheridan, Tickell, and some others also contributed; while those who superintended and arranged the whole,—the editors, one might call them,—were Dr. Lawrence, and one Richardson, a literary man about town, well-known in those days.

The principal writers of the poetry of the Anti-Jacobin were Canning, George Ellis, and Frere. The poem of " New Morality " was entirely by Canning, except the parody on Milton which was by Ellis.

October 27th.

News of the taking of Delhi by our army under General Wilson. The fighting seems to have lasted six days, from the 14th to the 20th of September. Our loss in officers heavy ; 50 officers said to have been killed or wounded : but of non-commissioned officers and men only about 600, which is not heavy for so long a struggle.

* Afterwards Lord Colborne.

The Richard Napiers left us yesterday, having come on the 26th ; I could not easily name, among my acquaintances, another couple so agreeable, or so worthy of love and veneration. Richard Napier is a remarkable combination of a powerful mind and extensive knowledge with a feminine refinement and delicacy of feeling. Indeed his sensibility is almost morbid, so that I fear he is not as happy as a man so excellent, so unselfish, so full of warm and tender affections ought to be. Although his favourite and especial studies are quite different to mine, I always feel myself thoroughly at my ease in his society, for he has one of those enlarged and liberal minds which despise no branch of study, but take pleasure in discussing and acquiring ideas in all arts and science. He is fond of argument, but is the fairest and most candid arguer I ever met with ; never out of temper, never overbearing, never sophistical, never arguing merely for victory, but always for truth. His wife is not less admirable than himself : with rare abilities, with very extensive and sound learning, with a disposition as true, as generous, and as affectionate as her husband's : but with more calmness and self-control, and a less excessive sensibility. I have known them almost as long as I can remember anything, and they have always been most warm and true friends to me.

LETTERS.

My Dear Katharine,

1857.   I thank you much for your kind and
agreeable letter of the 18th of October.   Mary's
letter, which you kindly sent us yesterday, was very
interesting, particularly to me, as I perfectly well
remember all the statues she mentions, and agree
with her about them, but I am rather surprised she
does not mention the Venus of Capua, which is very
beautiful, rather in the style of the Venus of
Milo, having like that, a dignity and grandeur
of character unlike what is usually seen in
a Venus.   The sitting Mercury which she mentions,
is one of the most beautiful things I ever saw.   I
entirely agree with her as to the mendicant habits of
the Neapolitans, and the abominably shabby
management of the Museum.   Everything at
Naples, in fact is an ideal of bad government, and
the face of nature is the only thing that its rulers
have not succeeded in spoiling.

I am afraid you are too sanguine in expecting
that the mutiny in India will *soon* be completely
quelled.   The capture of Delhi is indeed a heavy
blow, and a great discouragement to the rebels : but
their king has escaped, and as the place was not
invested by our force, I am afraid that most of the
villains will also have escaped, and the war will go on
though in a more scattered and desultory way than

before. However as the rebels could not succeed in destroying our handfuls of brave men before they were supported, they will have little chance after the reinforcements arrive ; but will they not carry on a sort of Pindarric war ? I expect that the heroism of our officers and soldiers will in the end save our Indian Empire ; but I do not expect the struggle will be a short one. Depend upon it, in such a crisis there is nothing like a military Dictatorship. General Wilson's orders to his army before the assault, I thought excellent.

Every care ought to be taken to search out those individuals who have exerted themselves to save and protect Europeans, and they ought to be most liberally rewarded. So also everything ought to be done to mark our gratitude to those native princes and chiefs who have stood by us in this emergency.

As to the task which is before our Government, when the mutiny shall have been suppressed —that of constructing a good Government for India—it is one of immense difficulty, but there cannot be a more important or more necessary one.

I find in Sir Thomas Munro's Life, which I am reading, that that eminent man was rather averse to the opening of the trade between England and India, at least he thought it a hazardous experiment, he thought too that from the habits and peculiarities of the people of India, it was not likely that the exports of our manufactures to that country would ever be very great. I should like to know, perhaps

1857. you can tell me from what you may have learned in India, how this has turned out.

I find that the Ferns in my collection are about 630 in number; I do not say all good species, but what are commonly admitted as such. Of these 246 are from South America (excluding Patagonia and Fuegia) and the West Indies. 136 from India and the Indian Islands (from yourself and Dr. Hooker); 80 from New Zealand (from Hooker); only 42 from Europe; and only 27 from the United States.

Those from other countries I have not yet counted.

Those that I have myself gathered in a wild state are 157. There are 5 or 6 of which I have specimens *both* from South America and India:—Asplenium cuneatum, Trichomanes radicans, Trichomanes rigidum, Gleichenia dichotoma, Gleichenia pubescens and perhaps Lygodium volubile; but the number of species common to the two countries is really considerably greater than this. I wish I could say that my knowledge was in proportion to my collection, but with respect to genera at least, I have hardly been able to arrive at any more definite notions than when I began; perhaps indeed less, for it is more easy to form large and decided conclusions when one has but a partial knowledge of a subject of this sort, than when one has a more extensive knowledge of its details.

Fanny is well, and sends love. My Father is going on well, and we have comfortable accounts of Cecilia.

The weather here is excessively damp, and I dare 1857 say it is not much better in London.

With love to your husband and children,

I am ever,

Your very affectionate Brother,

C. J. F. BUNBURY.

JOURNAL.

November 12th.

Fresh news from India, and very important. Delhi completely in our hands ; the old wretch of a king (said to be 90 years old) a prisoner : two of his sons and a grandson taken and shot immediately— quite right. But our loss has been very heavy, much more so than at first reported, 61 officers and about 1,180 men killed or wounded : about one-third of the force employed in the assault : and what is worst, General Nicholson has died of his wounds.—Lucknow has been relieved (*i. e.* the Europeans besieged in the Residency have been so) by Havelock and Outram, and part of the city taken, but this also with heavy loss on our side. General Neill killed. He and Nicholson are great losses.

November 16th.

More particulars of the capture of Delhi : a most splendid achievement, one of the finest in our military history : and equally honourable to General

1857. and soldiers.   But I believe we are yet very far from
the end of the mutiny.

———

November 16th—20th.

Our house full : a glorious merry madcap party,—
four Pellews, Edward, George, and William Napier,
and Fred Freeman.

———

November 28th.

Came up to London and established ourselves in
Charles Lyell's house, 53 Harley Street, which he
had kindly lent us.

———

November 29th.

Sunday.   Called on Edward, and had a good talk
with him.   He speaks highly of Mr. Buckle's book,
the " History of Civilization in England," which he
is reading, says that it is excessively theoretical and
dogmatical, but that the farther he proceeds in it
the more he is struck with the learning and research,
the earnestness and the ability with which it is
written.   He recommends also Daubeny's new book
on the " Husbandry of the Ancient Romans."   He
told me that the letters on Indian affairs, in the
Times, signed " Indophilus," were written by Sir
Charles Trevelyan.   The positive announcement by
the *Times*, that Ministers intend forthwith to
propose the taking away the government of India from
the Company, has occasioned much surprise and
remark.   Edward says he has little doubt that the
announcement is correct, but that it was not

intended to be so soon made public. We dined 1857. with the Horners, a pleasant family party : only Susan not well. Harry Lyell does not think there is any reason to apprehend that a famine in Upper India will ensue from the war : he thinks it will not have generally interfered with the cultivation of the soil.

December 1st.

Much milder but dismally dark.

Spent much of the day at the Athenæum, reading and in particular studying Asiatic researches. The style of Sir W. Jones's writings strikes me as stiff, heavy, and excessively artificial. In the third volume there is the very copious and satisfactory account of the great battle of Paniput (translated from the original of Casi Raja Pundit, who was an eye witness of the battle), of which Elphinstone has made use in his " History of India," and Fraser in the " Memoirs of Colonel Skinner." The style of this is remarkably clear and simple, quite free from oriental bombast and verbiage,

An agreeable little evening party at Mr. Horner's. I had pleasant talk with Lady Bell, Miss Moore, Professor Rogers (the American, Henry Rogers, lately appointed Professor at Glasgow), and Mr. Pulszky. Lady Bell much fascinated by Mr. Buckle's book, which she is reading.

I learned much from Mr. Pulszky about the Sanscrit language and literature, in the study of which he has been for some time deeply engaged.

1857. He says that the grammar of the Sanscrit is very
complicated and highly artificial; that it was
evidently much studied, for there are not only
very ancient special works upon it, but also large and
elaborate commentaries written in later times, on
those grammars themselves. There is in particular
one work on a very singular plan, a long poem
relating to the exploits and adventures of Rama,
but written expressly to illustrate the rules of these
ancient grammars, and serving as a commentary on
them ; this is done he says with great skill. The
language is extremely rich and copious, abounding
like German, with synonymous words. The
literature immense ; treatises on almost every
subject except history, of which there is hardly any.
As with the Alexandrian literature of Greece, a
large part of it consists of elaborate commentaries
on the earlier writings. There is a vast deal of
composition in verse of the most artificial and far-
fetched character—mere *tours de force* and strained
plays upon words.

———

December 2nd.

Dined at the Geological Club.

Dr. Livingstone was there as a guest, but did not
talk much. I had some good talk with Murchison,
Douglas Galton, and Colonel James of the Ordnance
Survey. Murchison asked me about the fossil
plants of the Rothliegendes or Lower Permian of
Germany, of which he said Göppert had sent him a
list, showing them to be with a few exceptions
(Neuropteris Loshii one of the exceptions) speci-

fically distinct from those of the coal formation, 1857.
though of the same genera. (But I am not sure
whether Brongniart's genus, Callipteris is found in
the true coal). He said that Beyrich is he believes,
the only geologist who refers the Rothliegendes to
the Carboniferous period. He told me that Pander,
a very accurate and painstaking Russian naturalist,
has found reason to melt down fourteen of Agassiz's
genera of fossil fishes into one.

December 3rd.

Very mild.

Meeting of Parliament. Read the Queen's
Speech as soon as it came out, at the Athenæum; I
like the way in which it speaks of the exploits of
our people in India. We visited old Mr. Carrick
Moore (now ninety-five years old), my cousin Bessy
Arran and Mr. William Grey.

Dined with the Henry Lyells; met the Charles
Youngs and Erasmus Darwin.

December 4th.

A fine day.

To the Linnean Society, and saw for the first
time their fine new apartments in Burlington House,
where they are now installed side by side with the
Royal and the Chemical Societies. The library is
spacious and handsome, though not quite so well
lighted as might be wished. Mr. Bennet whom I
met here, told me that the Society has lately
received a valuable present of the entire scientific

1857. correspondence of Sir James Smith, arranged and
bound, presented by his widow.
Edward dined with us.

* * *

December 5th.

We went down to Sandgate; the weather very
fine.

* * *

December 6th.

Sunday.   At Sandgate.   Spent the day very
comfortably with Henry and Cecilia and their lovely
little children.   Walked with Henry to the so-
called Cæsar's Camp; probably a British en-
campment.

* * *

December 7th.

Returned to London, and dined with Mr. Horner.
Very bad weather.

* * *

December 8th.

A horrible fog the whole day, so thick and opaque
that we literally could not see across the street.
Did not go out.—This fog was the thickest I ever
saw, and seems to have extended pretty uniformly
over all London, and a good way round; but towards
7 p.m. it cleared off suddenly, and the night was
rather fine.

* * *

December 9th.

Shopping with Fanny.   Dined with the Charles
Youngs.   The last news from India not com-

fortable. We must hope that the reinforcements 1857.
will reach Lucknow in time, but it will be a near
thing.

Susan Horner and George Napier dined with us,
very pleasant.

Fresh news from India, but not entirely satisfactory; Sir Colin Campbell had indeed reached
Cawnpore, and marched again from thence on the
9th November, for Lucknow: and Colonel
Greathed's column had preceded him by some
days, but nothing positively known of the situation
of the brave men either in the Lucknow Residency
or in the Allum Bagh.

Dined with my Uncle and Aunt at Clapham
Park. Sir William very much bent and crippled,
yet in good spirits and talking in great animation.
He is not, however, more cheerful in his views of
public affairs, nor more favourable in his judgment
of public men, than usual.—Talking of Raikes's
Diary, and of some instances of *prophecy* recorded
in it, Sir William mentioned the celebrated prophecy
addressed to Josephine, when very young, by a
fortune-teller in the West Indies,—to the effect that
she would be greater than a queen, but would fall
from her greatness and die in a hospital. (Josephine
died at Malmaison, which had originally been a
hospital). Sir William heard this told when he was
a boy some time before 1800, by Lady Ancram,

1857. who had been with Josephine when the prediction
was delivered.   At the time when my Uncle heard
it, Napoleon was not yet Emperor.

December 12th.

Foggy.

Dined with Erasmus Darwin; an uncommonly
pleasant party; the William Greys; the Herman
Merivales; the Wedgewoods; Mr. Newman,
("Phases of Faith") and Mr. Ferguson, the architect
(and the author also, I believe, of the new theory of
fortification.   It happened curiously enough that I
had this very morning at the Athenæum been
looking through Mr. Ferguson's beautiful work on
the " Architecture of India."   The talk was very
good.   I talked much with Mrs. Grey, who is
extremely pleasant; with Mrs. Wedgewood, who is
sensible, good-natured and cheerful; and with Mrs.
Merivale, who is lively and entertaining.

Mr. Newman's conversation is very good, he
seems to have great and various knowledge but
is modest and quiet; his appearance odd and
foreign, and though he has not exactly a foreign
accent, there is something foreign in his way of
speaking.   Mr. Ferguson seems to be a man of
great knowledge and ability.   He said that the
Palace at Delhi is a building of great solidity, and
he understands that it has not suffered much in the
siege.   It was built by Shah Jehan, little more than
two hundred years ago.

Delhi altogether, as it at present exists, is
altogether a modern city, but there is a vast extent

of ruins outside of it. He talked of the mud forts in 1857. some of the north-west parts of India, which are very difficult to attack, as they cannot be battered, for a cannon ball goes clean through the wall, cutting a round piece neatly out, without shaking the rest while the defenders stand on a bamboo stage within and fire over the top of the wall.

He mentioned also that in the Himalaya it is common for houses and even whole villages to be struck and set on fire by lightning; whereas in the plains of India, though thunderstorms happen daily at certain seasons, he did not remember to have heard of any accidents caused by them.

Merivale said that the district in Europe most noted for destructive thunder-storms and for accidents by lightning, is the country round Padua.

There was some talk about Mr. Buckle's book. Mr. Newman said it was very interesting, but he found some things in it very repugnant to his sentiments and opinions; he thought the style too diffuse.

Darwin said it was written in a clear, flowing, easy style, like that of the French, with perhaps too little compression.

December 14th.

Sally dined with us, and Edward came in the evening. He is sanguine about Lucknow being relieved in time. He thinks as I do, and I think it is the opinion of most who have attended at all to Indian affairs, that the man who more than any other has contributed to save the British Empire in India in this crisis, is Sir John Lawrence.

4 F

1857.   Edward told us of an amusing retort of Layard to
Mr. Bowyer,* the Papist Member for Dundalk.
They were discussing the condition of the Papal
States ; Layard, who was just returned from a tour
in Italy, inveighing against the mis-government of
those countries, the other defending their Government.   At last Mr. Bowyer said, banteringly—

" Ah you have those Nineveh bulls so continually
before your eyes, you cannot see anything else well."

To which Mr. Layard replied—

" Well, if the Nineveh bulls interfere with my
vision, certainly the Papal bulls interfere with
yours ! "

December 15th.

To Marlborough House to see the Turner
Gallery, now arranged in a series of rooms upstairs,
where the light is much better than on the ground
floor ; but though this was a comparatively fine day,
still many of the pictures were but very ill seen.
Most of the rooms are rather too small for pictures
which, like so many of Turner's, ought to be seen at
some distance.   On the whole, his water-colour
drawings pleased me most, especially those in sepia,
the originals of the Liber Studiorum.

We dined at Mr. Horner's.   Poor Clarke, who
was the husband of that most admirable of human
beings, Sarah Napier, is very ill.

December 16th.

I saw my dear old friend, Lady Napier, Sir

* Sir George Bowyer.

George's widow. We had not met since the death 1857. of that excellent and delightful person her husband. It was an agitating meeting. She still has much beauty, and all her warmth and sweetness of manner.

Fanny and I paid some pleasant visits—to Mr. Babbage, Lady Bouchier and my cousin Pamela.

Dined at the Geological Club: Mr. Robert Stephenson, the engineer, in the chair:—sat between him and Austen; Professor Rogers opposite to me. Some good talk. Inquiries about the severe accidents which Mr. Rogers' Brother met with while travelling by the railway to Norwich, led to mention of other remarkable accidents. Mr. Stephenson spoke of a curious one which happened some years ago, I think on the Great Western line: two trains were passing each other going opposite ways, when the *tire* of a wheel of one of the carriages broke, and the fragments flew off with such force, that entering a carriage of the other train, they killed two persons sitting in it. Austen mentioned that once when he and his wife were travelling by railway, the window of the carriage in which they were sitting was shattered to pieces by a pebble which had been thrown off by the passing of the train, and striking against a wall had rebounded against the carriage.

Professor Rogers mentioned the extensive deposits of tertiary coal on the West coast of North America; this led to talk of the climate of the coal period: and he said that the climate of the Oregon coast at the present day is somewhat of that

1857. character, — excessively moist, equable and mild.
He said (and this was quite new to me, and very
curious) that he believes this peculiarity of the climate
of the West coast, and its great difference from that
of the East to be entirely owing to the "Pacific Gulf
Stream," or great Chinese and Japanese current,
which flowing obliquely across the Northern Pacific,
strikes on the American coast to the North of the
Californian peninsula, and has exactly the same
effect on the climate of that coast, which the
Atlantic Gulf Stream has on the climate of Western
Europe.

Talking of the difficulty of referring the scattered
portions of fossil plants to their proper places,
Austen said very well, that some of the principal
geological difficulties were like parochial difficulties
—questions of *settlement*.

The weather still continues remarkably mild.
Mr. Stephenson said that the thermometer in
London, has not yet, at any time this winter, been
below 40 deg.

December 17th.

The newspapers report another and seemingly
most complete failure in the attempt to launch the
great ship, the "Leviathan." All the machinery
has broken in pieces, the ship sticks fast, and farther
proceedings seem to be adjourned *sine die*. I heard
yesterday at the Geological Club, that the cost of
these unsuccessful attempts to launch her has
already amounted to £60,000. Lady Bell, whom
we visited to-day, told me that the Bishop of

Oxford, meeting Mr. Brunel in society, said to him : 1857.
" Thinkest thou to draw *Leviathan* with a hook ?"
We called also on Mr. Boxall, and saw some good
portraits ; in particular a very spirited one of Miss
Hosmer, the young American lady who has shown
so much genius for sculpture. Boxall is an
accomplished and agreeable man, but valetudinarian
and hypochondriacal.

Went to the Linnean Society meeting. The
first paper read was on the Birds of New Guinea,
by Mr. Sclater ; showing the Zoology of that great
island, as far as yet known, to be more allied to that
of Australia than of the Indian islands. Next, an
important letter from Dr. Ferdinand Mueller, on the
botany of Tropical Australia, particularly of the
north-west part, showing the botanical results of an
extensive exploring expedition in which Dr. Mueller
has been engaged.

J. Hooker spoke of the value of the collections
made (which are now at Kew) and the accuracy of
Mueller's researches. The most curious fact is the
discovery of a new species of Adansonia or Baobab ;
the African one having hitherto been the only known
representative of the genus. The Leguminosæ are
just as in extratropical Australia, the most numerous
family of plants, and the Myrtaceæ (if I heard
rightly) next ; the Compositæ coming only third in
order of number. Acacia and Eucalyptus the two
most numerous genera. Hooker remarked that in
looking over the collections, he had been struck
with the resemblance of the botany of north-west
Australia to that of the Deccan and Mysore.

December 18th.

Fanny and I went to the British Museum, and spent nearly two hours among the Elgin and Townley marbles, principally indeed the former, studying with care those noble works of art. They are now very well arranged (the statues from the pediments I mean), and in a fine and well-lighted room. The Townley marbles also are now much better arranged than when I last saw them, and in a very handsome suite of rooms.

December 21st.

Dined with the Horners to celebrate Mrs. Horner's 71st birthday. The party—Mr. and Mrs. Horner, Fanny and I, the Henry Lyells, Susan and Joanna, Robert Brown (who was 85 years old this very day), Lady Bell, Joseph Hooker, Erasmus Darwin and Edward. Passed a very pleasant evening, with much good and cheerful talk.

Poor Thomas Clarke* died this morning.

December 22nd.

Went with Fanny to the National Gallery, where we admired the two noble pictures by Turner, and the New Paolo Veronese ("Alexander and the Family of Darius,") which is indeed a splendid example of that gorgeous painter.

We had a very pleasant little party to dine with us: my Cousin Caroline, Katharine and Joanna, Dr. Hooker, William Winthrop, and George Napier.

* Husband of Sarah Napier.

After dinner we all went to hear Albert Smith : very 1857.
comical and diverting indeed.

———

Went to the Museum of Practical Geology, and
examined the instructive models of Etna and
Vesuvius. There is a splendid mass of aggregated
crystals of specular iron, from Ascension Island,
resembling what is found at Stromboli, but the
crystals larger, and aggregated into a larger mass
than any I have seen from Stromboli. The
specimens of malachite from the copper mines of
South Australia are very large and very beautiful.

At the Athenæum, read in the newspapers the
telegraphic dispatch just arrived from India ;
Lucknow relieved by Sir Colin Campbell, after six
day's hard fighting ; the women and children, sick
and wounded safely removed to Cawnpore. This is
good news indeed, and we may well be thankful for
it. It is clear indeed that the war is by no means
over yet, but I trust that the worst is over, and that
what remains will be merely the effectually treading
out the embers of the rebellion. The great point
now is to get hold of the arch-scoundrel Nana
Sahib.

———

Went to a children's party at Mr. Pulsky's.
Here I saw Kossuth for the first time : a younger
looking man than I had expected to see, and less
care-worn in appearance : a mild, quiet, grave,
thoughtful face, not handsome, but with a very good

1857. expression. — There was one very pretty woman among the company—a Madame Roche, daughter of the musician Moscheles : of Jewish blood I understand.

_____

December 25th.

Dined with the Henry Lyells : a pleasant family party, with the addition of Lady Bell and Edward. Edward gave us a curious account of an American who has lately come to England, having (or professing to have) a secret by which he can immediately tame and acquire an absolute power over the wildest and most unruly horses. Edward had his information from Kinglake (" Eothen ") who is very intimate with, and the legal adviser of General Airey : and Airey is the man with whom the American has been principally in communication on the subject. General Airey has in fact been admitted into the secret, but under a bond to a large amount not to divulge it. He is said to be quite satisfied of the value of the method ; and Kinglake, who was present at some of the experiments, said that the results which the man exhibited were certainly most remarkable. It is not like the case of the famous Irishman (" the Whisperer ") who was said to be able to subdue the most vicious horses : _he_ pretended to work by a peculiar " charm," or faculty, which he could neither explain nor communicate ; but this man (who was a horsebreaker by trade) merely professes to have a secret, by which anybody, when once in possession of it, can produce the same effect ; but naturally enough he wishes to enrich himself by this secret.

We left London, travelled by the railway to Bristol, and thence by carriage to the MacMurdos house at Westbury, where we were most kindly received.

———

December 27th.

Sunday. Began to read Mr. Buckle's "History of Civilization." Took a pleasant walk with Mac Murdo to Durdham Down and to the edge of the cliffs overlooking the Avon, near Cook's Folly: but the day though mild and fine, was too foggy to allow of our seeing the beautiful scenery to advantage.

———

December 28th.

Went with Susan MacMurdo to visit the Mileses at King's Weston. Lord Arran and Bessy are staying with them at present. King's Weston park appears to be beautiful, but the day was very foggy and we could not see far. The house is fine and stately : built by Vanbrugh ; it belonged formerly to the family of De Clifford. A noble hall of surprising height, hung round with old family portraits. The gardens extensive and fine, with much glass. A standard Magnolia grandiflora, unprotected in the open ground, large and vigorous, is an indication of the mildness of the climate. We observed also Woodwardia radicans growing in the open air. In the hot-houses we saw some beautiful Ferns and Lycopodiums : the Poinsettia in full flower : the Meyenia (near to Thunbergia) very handsome : the

1857. Thyrsacanthus, a pretty shrub with long thread-like weeping branches, and scarlet flowers, and a fine Franciscea, known as the Paraguay Jasmine, and introduced to England by Lord Arran,—it has large sweet scented flowers, of a fine violet colour when they first expand, and gradually changing to white.

A pleasant little dinner party at the MacMurdos; the company—the Dowager Lady Napier, and her daughter, Mrs. Hope, both very pleasing.

December 30th.

We went with MacMurdo to see the barracks of the military train at Horfield. The arrangements appear to be exceedingly good, and I was particularly pleased with the neat and ingenious packing of the waggons—the happy contrivances by which each waggon is made to carry outside of it, every article necessary for keeping the waggon itself and the team in the most perfect state of efficiency; while the interior remains perfectly clear and free for the baggage of troops, or whatever else it may be required to carry.

In the evening to King's Weston, where capital private theatricals were performed, followed by dancing.

LETTERS.

From his Father.

Folkestone,
January 30th, 1858.

A great many thanks to you my dear Charles 1858. for your very agreeable letter from Oaklands. You are such a fly-about couple that I have difficulty in sending answers. I am very glad that you have met Lady Dalrymple* and have found her well. I entertain a very sincere friendship for her.

Indian affairs must still give us long anxiety. What a mercy it was that Government had the good sense to dispatch Sir Colin Campbell at once. His relief of Lucknow with so small a force as that which he actually had under his command, was one of the finest pieces of Generalship that I remember.

The protracted war is forming good officers in the British army ; let us hope that it may not produce a Hyder Ally on the other side. A great and continual supply of troops from England will be indispensable, and yet Government have, through a stupid notion of economy, let slip the most favorable time for getting recruits. Lord Panmure is almost as intolerable as Lord Clanricade.

The opposition to the India Bill is likely to be so formidable that Palmerston may feel it expedient just to lay it on the table, and let it die a natural

---

* Lady Dalrymple aud her husband were great friends of Sir Henry and his first wife, at Messina, in the early days of their marriage.

1858. death, or that if he shonld be confident enough to press it forward, he may be turned out before Easter.

I agree with you in thinking that the proceeding has been premature, to which I add that the announcement of so sweeping a measure has been rash. I should be inclined to establish a Dictatorship of India for a couple of years, and during that time the details of the great change might be matured. But where is the man fit to be Dictator? Sir John Lawrence?

Before I was taken ill last June, I had ordered a monumental tablet to be made with the intention of having it put up in the Church at Barton, as sacred to the memory of your dear Mother. Just as I was leaving home I received notice that the tablet was finished, but it still remains at the sculptor's. I wish very much that you would go and see it, and let me know if it has been well executed.

Much love to Fanny. to whom I owe a letter.

Most affectionately yours.
H.E.B.

West Ashling,*
February 4th, 1858.

My Dear Mrs. Horner,

I thank you heartily for you affectionate note and kind wishes on my birthday. I am in very tolerable preservation, thank you, and not conscious of being much aged since yesterday, nor indeed since this time twelvemonth. It is a great comfort to hear such good accounts of you and of all our dear friends

---

The house of his cousin, Gen. William Napier.

in London ; and my Father writes quite cheerfully 1858. from Folkestone.

We had a heavy fall of snow here the day before yesterday, but it was all gone by the next morning, and to day we have mild rain.

I hope Mr. Horner will not have caught cold at the Geological Society. I reckon much on the pleasure of hearing his Nile paper read at the Royal Society, while we are in London.

My chief occupation here is reading " Buckle," which I find very interesting and which improves much as I go on. It is not a book to read fast, for there is matter for thought in every page.

I always thought (I mean since I have been reading it) that it was a book that would please Mr. Horner, and Susan writes me word that he is delighted with it. I dislike many of Mr. Buckle's views and habits of thought, especially his excessive dogmatism, but I admire not only his learning and ability, but his hearty love of freedom and of general improvement.

The marriage of the Princess* was indeed very interesting ; it is not often that one can sympathize so thoroughly with the parties to a Royal wedding.

We hope to be with you on Tuesday next, the 9th and to stay a few days with you and a few days with the Lyells, before returning to Suffolk.

Pray give my love to Mr. Horner and to Susan and Joanna, and many thanks to the two latter for

* The Crown Princess of England with Prince Frederick of Prussia.

1858. their very kind letters ; also to Katharine ; I will
write to them very soon.

Ever your very affectionate Son-in-law,
C. J. F. BUNBURY.

West Ashling,
February 4th, 1858.

My Dear Mary,

Very many thanks for your kind letter and
kind wishes on my birthday. You and I are, as you
say, embarked in the same boat, so far as having
been born near the same time ; but if I were nearly
as young for my age as you are, I should have reason
to be very well pleased. However, I do not at all
complain or repine at the march of time. I do not
think I ever was particularly young, and I am
inclined to agree with Thackeray as to the many
advantages of being an old fogey. I am thankful to
say that time has not diminished my power of
enjoying nature, or books, or the society of dear
friends, and I have reason to be doubly thankful for
the happiness I enjoy in my home. I am growing
very prosy—a symptom of age by the way—but
I really have marvellous little to tell you. Our visits
to Oaklands* and Hamblecliff† were very enjoyable. Minnie was very pleasant indeed, Sarah most
bewitching, and I had great pleasure also in meeting
my Mother's old friend Lady Dalrymple, a very
agreeable person. Augusta very cordial and pleasant,
as also her Husband, and my dear old friend Lady

---

* Lady Napier and her daughter, Mrs. John Napier.

† Mr. and Mrs. Frederick Williams Freeman.

Napier (Sir George's widow), most affectionate 1858.
to us.

I have finished up my Paper on the Madeira fossil leaves, for the Geological Society, and am now engaged on an essay which I am to read some time or other at the Athenæum at Bury; the subject, —"Notices of some of the most eminent English Naturalists, as illustrating the rise and progress of the Study of Natural History;" but I cannot do much in it here, at a distance from all books of reference. I hope we shall be in London before the middle of next week, and shall be very happy to pass a few days with you in Harley Street, as you have kindly asked us.

Much love to Charles Lyell, who I suppose is very busy with Etna.

Ever your very affectionate Brother,

C. J. F. Bunbury.

West Ashling,
February 5th. 1858.

My Dear Katharine,

Many thanks for your kind letter and good wishes on my birthday. Though the country hereabouts is pretty, there are no particularly interesting objects; except indeed a remarkable grove of enormous Yew trees, in a hollow of the chalk downs called Kingley Bottom, about a couple of miles from here, a very pretty spot, to which I walk most days. The Yew trees are of prodigious size, and must be of great age; there is some legend or story about their having been a Druidical

1858. Grove, but I have no doubt they are indigenous. The slopes all round the hollow are covered with wild thickets of Yew and Juniper, and in summer they are very pretty botanizing ground.

On some of the trees I have found Pterogonium Smithii, not a common Moss in England; also Neckera pumila, Leskea complanata, and I believe Leskea trichomanoides. The common Ferns hereabouts are the Hartstongue and Polystichum aculeatum.

The accounts from India are not comfortable to my thinking, and do not look as if the war were near an end. Almost everything seems to depend on Sir Colin Campbell, indeed it is rather frightful to think how much depends on the life of one old man. He has shown himself a worthy disciple of Sir Charles Napier.

I hope we shall not have a war with France in addition to our Indian troubles, but the publication in the *Moniteur* of all these offensive and outrageous addresses, shows an evident inclination to bully us, which must be resisted. I should probably disagree much with you about the general questions as to the French Government and the conspirators, and I have an utter abhorrence of assassins, but it would never do to submit to intimidation. Altogether I do not think our prospects just now are very agreeable.

I hope to see you all next week. Love to Harry and your children.

Ever you affectionate Brother,

C. J. F. BUNBURY.

Mildenhall,
March 25th, 1858.

My Dear Katharine,

I am very glad that you have found additional interesting things among the Assam Ferns
that Mrs. Agnew sent you, and that you have now
and then leisure to attend to them. They are most
beautiful and interesting things, but most bewilder-
ingly difficult as to arrangement and grouping and
distinction ; and their arrangement is become a
perfect chaos, for no two botanists agree about it,
and the more there are who write upon the subject,
the greater the confusion seems to become. I wish
as much as you do that there were some book
giving a clear and concise view of the genera,
pointing out their affinities and distinctions ; but I
doubt whether such a thing be possible in the
present state of our knowledge. I doubt whether
Sir William Hooker himself (whom I look upon as
the very greatest authority on the subject), settled his
own views or formed any definite arrangement of the
genera ; indeed I think it is almost clear, from his
observations in the *Century of Ferns*, that he has not.
Indeed one has only to look through that book,
or to go through the descriptions of the different
sections of Davallia in the species Filicum, to see
how great the difficulties are. The *indusium* not
only differs in species evidently most nearly allied,
but it certainly, in some instances, varies in the
same species : but so do the veins likewise, and
as far as we can yet see, the anatomical structure
does not furnish any better characters,—I mean the

1858. internal structure of the stem or stalk. Those great
leading differences in the capsule, on which the
tribes of Schizæineæ, Osmundaceæ, and so forth are
founded, seem tolerably clear and constant, and to be
the only ones which can be relied upon. Then, as
to the species, it is not less difficult to determine
the limits of them, so great is their liability to vary.
One can hardly venture to say how many supposed
species may not ultimately be swallowed up in
Asplenium marinum and Asplenium Trichomanes.
It seems to be now quite clear that quite contrary to
what Humboldt formerly stated, Ferns have, as
a rule, a very wide geographical distribution, and
that a large number of them are common to the old
and new world.

I hope you will come to us in May, and then
we can discuss all these questions while looking over
my collections together, for they run to too great
a length for any letter. Have you seen Berkeley's
"Introduction to Cryptogamic Botany?" It contains
a prodigious deal of information, but is terribly ill
written and hard to understand.

We went to Barton on the 18th, stayed till the
23rd, then went to Ickworth for a day and returned
home yesterday.

My Father returned to Barton just before we left
it; he is looking very well. We spent a very
pleasant day with the Arthur Herveys; they are
people whom it does my heart and mind good to
associate with. We saw Lady Cullum, too, looking
remarkably well.

So the villainous old King of Delhi is to be sent

to the Andaman Islands. I hope the government at 1858.
Calcutta will not let him off. I see it is expected
that an amnesty will be offered to the people of
Oude on condition of their submitting,—with the
exclusion of the mutinous Sepoys and of those who
have actually been concerned in murders. This
is all right. Much as I am opposed to the plans
of general indulgence advocated by some, I think a
great distinction ought to be made between the
general population of Oude (who can hardly be
considered even as rebels, since their country was so
lately independent), and the murderous mutineers.
I wonder that no steps are taken to ascertain the
degree of guilt of the King of Oude or his Minister.

We are looking forward to the pleasure of seeing
Mr. and Mrs. Horner and Joanna about the 10th of
next month ; and I hope we shall not be disappointed of our promised visit from you and Harry
and your children, which I suppose will be in May.
I am very glad to hear so good an account of them
all. Fanny is well, in excellent good looks, and
busy as a bee.

> Ever your very affectionate Brother,
> C. J. F. BUNBURY.

JOURNAL.

April 8th.

Charles Lyell, in a letter to Mr. Horner, mentions
a joke that is current in London about Buckle ; the
author of the great book I lately read ; he is such

1858. an overwhelming talker, that "people say he makes them wish for *a Buckle without a tongue !*"

Mr. Horner when at Edinburgh lately, heard from Lord Minto the following anecdote of Louis Napoleon, which Lord Minto had from Sir Edward Lytton Bulwer himself :—

When Louis Napoleon was preparing for his attempt on Boulogne, he told Sir E. L. Bulwer that there were three things he was determined to do, if he succeeded in making himself master of France :—

    1st. To abolish the Constitution.

    2nd. To put down the liberty of the press ; and—

    3rd. *To invade England.*

April 9th.

News of the capture of Lucknow by our army under Sir Colin Campbell, with no great loss on our side ; but unfortunately most of the rebels seem to have escaped, our force not being sufficient to surround the place. I fear they will now scatter and keep up a desultory warfare and be more difficult to deal with than when they were concentrated.

April 18th.

News of the acquittal of Bernard, the French refugee, tried at the Central Criminal Court on the charge of murder, for complicity in the attempt to assassinate the Emperor of the French.

The three little Lyell boys, Leonard, Frank, and Arthur are uncommonly fine children, and of great promise, Leonard showing extraordinary powers of mind for his years.

April 23rd.

I gave my lecture on India ("The Rise of the British Empire in India"), before the Mildenhall Literary Institute.   I was much pleased by the way in which Katharine and Mr. Horner afterwards spoke of it.

April 24th.

Our dear friends the Horners and Henry Lyells left us this morning to my great regret ; their visit has been a most pleasant time to me.   I have had great satisfaction in going through my collection of Ferns with Katharine, who studies those plants enthusiastically and with great intelligence.

I read a very interesting manuscript which my Father had sent me : a military narrative of the expedition of St. Louis to Egypt, taken from Joinville, with many comments and criticisms of his own.   He wrote it about 1832.

April 27th.

To London with Fanny, to Charles Lyell's house in Harley Street.

Visited the French Exhibition of Pictures in Pall
Mall, but was not particularly struck with it. There
is a large Dog, by Rosa Bonheur, very good, but not
equal to some of Landseer's. Pictures in the *genre*
style predominate; there are none of those daring,
startling pictures of great or terrible subjects which
struck me at Paris as characteristic of the French
School, and which, though often exaggerated or
disagreeable, are impressive. The work of greatest
pretension, perhaps, is Ary Scheffer's "Margaret"
(from Faust), which I think cold, inanimate, and
insipid. I visited Burford's Panorama of Delhi; it
is not altogether satisfactory to me, the groups
of fighting figures being made too prominent, and
the buildings not sufficiently so.

Attended the meeting of the Geological Society
(for which indeed we had come to town). Read
my paper on the fossil leaves from Madeira, which
was very well received. Lyell gave a clear and
satisfactory account of the geological relations of
the bed.

———

Went with Lyell to the Botanic Garden in the
Regents Park. The profusion of flowers in the
"Winter Garden," or large greenhouse, is very
delightful; the Chinese Azaleas in exceeding beauty;
some Camellias also very beautiful, especially one
of a most exquisitely pure white. I noticed also
some fine trees of the Araucaria excelsa, and of
Rhododendron arboreum; thriving young plants of

Araucaria Bidwillii ; Grevillia robusta, tall and fine, 1858.
but not flowering ; Hakea Victoria, with curious
leaves ; and the very singular Aralia crassifolia.
There are some very good Ferns too : Woodwardia
radicans, very fine ; Hemitelia horrida, with the
trunk sufficiently developed to show the remarkable
leaf scars ; and Dicksonia antarctica.

Went afterwards to the Exhibition of the Society
of Female Artists ; but was on the whole rather
disappointed by it. Mrs. Murray's drawings are
excellent, particularly a house at Icod in Teneriffe,
and a group of market people at Rome.

Edward, with whom I walked back from the
Athenæum, gave me some account of Mr. Gladstone's new book on " Homer," which contains some
very singular things. Mr. Gladstone, he tells me,
discovers the doctrine of the Trinity, and most of
the mysteries of orthodox Christianity in Homer's
mythology.

April 30th.

We visited Minnie and Lady Napier, and had a
pleasant talk with them ; went with Minnie and
that most fascinating fairy, her daughter Sarah, to
Colnaghi's to see Sarah's portrait (a photograph)
which Minnie has kindly had done for us.

Read in *Fraser* an interesting Paper by Kingsley,
entitled " My Winter Garden." Most of Kingsley's
writings (the latter ones at least) are very attractive
to me. There is much with which I heartily
sympathize in his ways of viewing both mankind
and outward nature,—more especially the latter.

Weather very strong and cold. Fanny went to
Sandhurst.* I visited dear Katharine, and had a
very pleasant talk with her. Then looked into the
Zoological Gardens. The Honduras Turkey,
Meleagris ocellata, a splendidly coloured bird, looks
very thriving in spite of the cold: so do those still
more splendid creatures, the Lophophori or Monaul
pheasants of the Himalayas. In the Parrot house,
the Toucan is very lively, hopping with great
activity from branch to branch.

There is also a Hornbill, in very good condition:
and two specimens of a rare bird, a new acquisition,
the Podargus Cuvieri, of the goatsucker tribe, from
Tasmania. In general appearance, this Podargus
reminds us almost as much of an owl, as of a
goatsucker; one might fancy it intermediate between
the two. As might be expected of a nocturnal bird,
it appears sleepy and torpid in the day time.

At the Athenæum, I studied the botany of the
French Expedition, Scientifique de Morée; and
Gaudichaud on the botany of Freycinet's Voyage.

May 4th.

To Barton, where we found my Father very well,
on his 80th birthday.

May 5th.

My Father in talking with me of military men,
said that in his opinion, Sir Charles Napier was

* To the William Napiers.

superior in real military genius and military know- 1858.
ledge to all his contemporaries, even to the Duke of
Wellington himself; but he did not think that he
could ever have done what the Duke did, because
he wanted the patience so necessary under the trying
circumstances of the war. Nothing but the composed and steady patience of the Duke could have
carried him successfully through such obstacles and
impediments. Sir John Moore, my Father thought,
was not quite a *first*-rate General—would never have
accomplished very great things as a Commander-in-
chief, though he was admirable as second in command.

May 6th.

Returned home.

LETTERS.

Mildenhall,
May 12th, 1858.

My Dear Katharine,

I return *The Times*, having read Mr.
Russell's letter with very great interest; pray give
my hearty thanks to Harry for sending it to me.
Mr. Russell is certainly an excellent writer, and a
man of great ability; he gives one a most vivid idea
of the strange scenes he has witnessed, and there
seems to me much good sense in his observations.
It is very lamentable that our troops should plunder
so and make such wanton havoc in the fine palaces
of India: but I am afraid our armies have always
had a turn that way; the Duke of Wellington's
dispatches from Spain are full of complaints on the

1858. subject, and I am afraid our people plundered terribly at Badajoz and St Sebastian. But the French were still worse, and I dare say the Russians would be no better. As to what Major Nicholson says of the unmerciful severity of our soldiers, it is partly explained by one thing which Russell mentions ; that more of our officers have been killed or wounded by lurking enemies after the fighting was apparently over, than in the heat of action. This exasperates men, and indeed makes it imprudent for them to spare such insidious enemies. Again our men know that they are fighting against enemies enormously superior to them in number, and enemies too, who, if victorious, would assuredly not spare *them*. All this inevitably gives a ferocious character to a war.

I confess I cannot bring myself to feel very sorry for any amount of punishment inflicted on the mutineers. Cromwell, "our chief of men," was tolerably severe (under rather similar circumstances) to the Irish rebels, and I refer you to Carlyle's justification of him.

There is likely to be a pretty fight in both houses, to-morrow evening, on the question between Lord Ellenborough and Lord Canning ; I shall not be much surprised if the Government are beaten, and if they are they will probably dissolve Parliament.

In spite of the cold winds the Polypodium Dryopteris is showing its pretty fronds in our out-door Fernery ; Blechnum boreale and Nephrodium dilatatum also peeping. In our greenhouse, Adiantum formosum, Pteris hastata, Pteris rotundifolia, Osmunda regalis, and several others are 1858. expanding beautifully.

I hope your dear pretty children are quite well, and do not forget Mildenhall; the boys at least for we can hardly hope that Miss Rosamond will remember us long.

Pray give my love to them all, and to your husband, and believe me,

Your truly loving Brother,

C. J. F. BUNBURY.

JOURNAL.

May 16th.

Susan Horner sends us a very neat epigram (told to her by Dean Milman) which has been made on Mr. Hodge, who was suspected of concern in the plot to assassinate the French Emperor :—

> " What ? Hodge an assassin ?
> Oh no ! says his kin ;
> *Double ass.* if you will,
> But without any *sin.*"

LETTERS.

Mildenhall.

May 23rd, 1858.

My Dear Mr. Horner,

The weather is again showery and unsettled but very bright and pleasant between the showers, and our garden and paddock are in the full beauty of spring, the grass and leaves exquisitely green, and fresh flowers coming out every day ; Laburnums,

1858. Lilacs, Pæonies, Tulips, Irises, Abies, rejoicing the
eye in every direction.   I wish you and Mrs.
Horner and all the family party were here with us
to enjoy our paradise, which we have all to
ourselves ; almost like Adam and Eve ; but I do
not suppose Adam was ever troubled about police
cases or paupers.

Well, to be sure, the grand debate has come to a
curious termination ; the two armies marching off the
field at once with drums beating and colours flying ;
each claiming the victory, I suppose ; at least the
boasting of the respective newspapers is considerable.   I thought Roebucks speech excellent and
most conclusive ; Bright's very powerful, but as
usual with him, he overstated his case, especially
in the latter, and became needlessly offensive.
Roebuck was more logical.   I am not in general an
admirer of Sir James Graham, but I thought his
speech in this debate, admirable.   I differ indeed
from him in so far that I do not think the responsibility of Lord Ellenborough, can constitutionally
be separated from that of the whole Ministry.   The
correspondence between Sir James Outram and
Lord Canning is indeed important ; The Governor
General has certainly much to say for himself, yet
on the whole I think Sir James in the right ;
certainly not because I have any partiality to him.
I am very glad your paper on Egypt is printing, and
look forward with much pleasure to reading it.   The
discovery of the new bone cave in Devonshire is
interesting.

I have just begun De Tocqueville's book, your

birthday present to me "L'Ancien Regime et la 1858. Revolution."

In the evenings I read Scott's "Monastery" to Fanny; it is by no means one of the best of his novels, yet I am much struck in reading it, with his superiority in many respects to the generality of novelists.

With much love to Mrs. Horner and Susan, I am ever,

Your affectionate Son-in-law,

C. J. F. BUNBURY.

JOURNAL.

May 30th.

The fourteenth anniversary of our happy wedding day. Beautiful weather.

Finished reading Joseph Hooker's paper on Balanophoreæ, in the last part of the Linnean Society's Transactions; a most remarkable and masterly essay, beautifully illustrated, expounding in the fullest detail every part of the structure, and the affinities of a most singular and anomalous tribe of plants.

LETTERS.

Mildenhall,

June 2nd, 1858.

My dear Lyell,

Can you give me any information about the Equisetites that bears your name,—Mantell's Equisetum Lyellii, from the Wealden of Sussex?

1858. I have seen no specimens, but from the figures in Mantell's book on the British Museum, it would appear to be a true Equisetum. If you happen to have a tolerable specimen, and would lend it to me, I should be much obliged ; if not, I must just wait till I can visit the British Museum. I am working at the fossil Equiseta, a subject suggested to me by my having some some pretty good specimens of Equiseta infundibuliformis (which I think I showed you) from Cape Breton ; otherwise I find, when I try to go at all deep into the subject that my materials are not plentiful. But John Phillips* has been very good-natured in answering my queries about Equisetum columnaris, and giving me all the information he has collected about it,—which after all amounts to very little. It is odd enough (I was going to say singular, but it is by no means a singular case in fossil botany) that we should still have such imperfect knowledge of a plant which is found in so many different localities, and in very great abundance in some of them and which has been known above thirty years. Both its fructification and its internal structure seem quite unknown. Joseph Hooker, with his usual frankness and liberality has sent me copious information about the geographical distribution of the recent Equiseta, which I hope to bring to bear on the history of the fossil ones.

With respect to the Calamites, the more I look at that extraordinary specimen figured in the last edition of your Manual, the more it puzzles me to understand its structure. It is very incompre-

* The Geologist living at York.

hensible. But I suppose the accuracy of the 1858.
drawing may be depended on.

(June 8th).—I see your paper on Etna and
Vesuvius announced to be read at the Royal Society
next Thursday. I wish I could be there to hear;
but at any rate I hope to read it at some future
day.

We have glorious summer weather, which I dare
say you enjoy in the Zoological and Botanic Gardens; but though *they* may have a greater variety of
plants and greater rarities than we have, I hardly
think anything can be more beautiful or more
enjoyable than our garden has been for some time
past. I never before saw it so rich in flowers. The
blossoming of the Laburnum has been peculiarly
fine this year. Our garden is all alive with birds,
too, and this is a great pleasure to me. I delight to
watch them.

I had quite a surprise this morning : one of
Fanny's schoolboys brought her a quantity of wild
flowers which he had gathered in his walk, and
among them was a plant (Geum rivale) which I had
never yet seen in this neighbourhood. Mr. Saxton
is studying botany a little.

When do you go abroad ? Why cannot you and
dear Mary come and pay us a visit here before you
cross the seas ? It is a long time since you have
seen us in our summer dress, and I do assure you
we are much better worth seeing at this time than in
winter. I wish you would think of it. We are
looking forward to a visit from Minnie and Sarah
Napier this month, but hitherto we have been living

1858. in utter solitude since we came from London ;
it is indeed " a populous solitude of bees and birds,"
and unfortunately though the solitude is perfect as
far as society is concerned, I am by no means free
from interruptions, for the profane vulgar are
continually quarrelling and breaking the law ; and
I am the only magistrate now in the district.

Poor old Mr. Brown ! I suppose in a day or two we
shall hear of his death. It cannot be called un-
timely, but the departure of an eminent man must
always leave a void.

With much love to Mary,

Ever affectionately yours,

C. J. F. BUNBURY.

JOURNAL.

June 12th

A letter from Charles Lyell mentions the death of
our great botanical patriarch Robert Brown, and
gives some interesting particulars of his last
hours :—

"Yesterday morning Robert Brown breathed his
last. They—Brodie, Bright, and Boot—told him
they might keep him alive, even till Christmas
possibly, by opium and stimulants ; but he preferred
not to live with a mind impaired, and so, cheerfully
and tranquilly and in full possession of his intellect,
gave way to the break up of nature. Every one who
has been with him in his last days agrees with
me in admiring the resignation with which he

met his end, and the friendly way in which he talked and took leave of us all."

Received a panicle of the beautiful flowers of the Pavia Indica, which has blossomed at Barton for the first time probably in Europe. Sent a careful description of it to Joseph Hooker.

Minnie Napier and that most charming of children her daughter Sarah, came to stay with us; a great pleasure to both of us. Sarah is equally lovely and amiable.

We had a very pleasant day at Ely, with my dear friend Arthur Hervey, his excellent and agreeable wife, their two eldest daughters and Minnie and Sarah Napier.

The weather gloriously fine and very hot, as it has been for some time past. A good deal has been done in the restoration and adorning of Ely Cathedral since I was last there. In particular, the new "rere-dosse" or screen behind the altar is of wonderful beauty: it is of Derbyshire alabaster, most exquisitely sculptured in the richest and most elaborate Gothic style, with very numerous figures, and many small spirally wreathed columns (like those which one sees near the high altars of some of the old basilicas at Rome), inlaid with agate and

1858. gold. The beauty of the general effect is as remarkable as the exquisite finish and elaborate richness of
the details. The mosaic pavement before the altar
is also of great beauty.

_____

June 27th.

Heard of the death of old Dawson Turner; I do
not know exactly what was his age, eighty-three
according to the newspapers, but he must have been
much above 80. Since the death of his first wife,
he has been living in complete retirement. He was
certainly a man of very considerable ability, and in
his younger days did important and lasting service
to botanical science by his excellent books on Fuci
and Mosses. Much interest also attaches to him as
one of the earliest members of the Linnean Society,
as the intimate friend of Sir James Smith, and one
of that zealous band of botanists, whose labours
established on so firm a basis our knowledge of the
vegetation of our own country.

Louis Mallet, who is staying with us at present
(he came on the 24th), has told us one or two
anecdotes which are worth noting down :—

1.—A saying of Lady Theresa Lewis, respecting
Gladstone :—

Some one asked her whether he was likely to join
the Derby Ministry; "No," she said, "not just
now; at present all his energies are concentrated
on the endeavour to persuade Mrs. Gladstone to
call upon Helen."

This alludes to his curious book on Homer, and

his strenuous endeavour to vindicate Helen's 1858. character.

2.—The same Lady Theresa said of the Peelite party :—

" They are always putting themselves up to auction and buying themselves in."

3.—The Duchess of Manchester's appointment to be Mistress of the Robes, originated in a mere joke.

Before the Conspiracy question came on, and while Lord Palmerston seemed most firmly established in power, the Duchess, meeting Lord Derby in society, rather mockingly asked him *when* he was coming into office ; he replied, jokingly, that whenever he did come in, she should be Mistress of the Robes. When Lord Palmerston was so unexpectedly overthrown, and Lord Derby, as unexpectedly to himself as to every one else, actually came into power, the Duchess claimed and insisted on the fulfilment of his promise.

July 1st.

Our pleasant party broke up : dear Minnie and Sarah leaving us for Lord Raleigh's and the Pellews for Bury ; Louis Mallet having left us the day before.

A most pleasant time we have had with them. Louis Mallet is very agreeable, and seems very good ; Minnie charming.

July 2nd.

Heard of the death of old Mrs. Marcet, a superior

1858. woman, and a very useful one in her time; but she had long outlived her faculties, and for years past had been only a burden to herself and her friends. What a miserable condition!

LETTERS.

Mildenhall,
July 5th, 1858.

My Dear Katharine,

I must write you a few lines before you leave England, to wish you a good journey, and health and enjoyment during your stay in the wilds of Germany.

Pray dear Katharine take care of yourself, and avoid fatigue.

You will have heard, or will soon hear, of a little love romance, which has interested Fanny and me very much, and which has just come to its *denouement*—I trust happily for all parties*.

How fortunate was Robert Brown to die not only full of years and honour, but in full possession of his faculties and feeling that he was generally regretted.

I have not done much in Ferns lately, but have been chiefly occupied with my South Brazilian and Buenos Ayrian plants, working out the Compositæ, which are very numerous, and a difficult and troublesome family. They are very numerous at

* The engagement of Mr. Louis Mallet to Miss Fanny Pellew.

the Cape too, but chiefly of different genera, and even different subtribes. I have just finished reading Piazzi Smyth's Teneriffe. I am now going to begin the 4th volume of Cosmos.

We had a delightful visit from Minnie and Sarah, and now are hoping soon to see Minnie back again : and looking forward too to a visit (though I fear a very short one) from Charles and Mary. God bless you dear Katharine ; much love to your husband and your lovely children,

Ever your very affectionate Brother,

C. J. F. BUNBURY.

Mildenhall,
July 6th, 1858.

My Dear Joanna.

I thank you very much for your kind and interesting letter of the 1st from Berlin, as well as for the little photograph of Humboldt. I am very glad to hear that your expedition to Copenhagen was so prosperous, and your voyages so agreeable, and I was much interested by what you tell me of Copenhagen and its neighbourhood of which I knew very little before. Professor Leibmann is a loss to the botanical world (he must have been a young man); I know Professor Oersteds name as a botanist,—he travelled in Central America, and a paper of his on some of the plants of that country is translated in Sir William Hooker's Journal of Botany Denmark has produced some very eminent botanists, though none quite of the class of Linnæus and Jussieu. I remember Edward speaking of the

1858. abundance of superb Beech trees in Denmark : and
he said that the scenery of the Sound reminded him
of the Southampton Water,—very unlike the ideas
suggested by Campbell's " Wild and stormy steep,
Elsinore."

Robert Brown's death-bed seems to have been a
model of a calm, cheerful, philosophical close of life,
such as one might wish for one's self and one's
friends,—living to an advanced age, and then dying
surrounded by his friends and in full possession of
his faculties.

I hear that Mr. Bennett has been appointed Mr.
Brown's successor in the British Museum ; very
properly I think : he is a very good botanist and a
very amiable man.

The Pellews have been here, and Louis Mallet
and the Arthur Herveys for a couple of days, and
Sarah was very happy with the Hervey girls, and
with Georgy Pellew. And now Fanny Pellew and
Louis Mallet are engaged to be married, and I
think it is a very good match, and I am very glad to
find that all parties, Mr. Mallet and all, are so
much pleased. I like Louis Mallet very much, and
Fanny Pellew is an uncommonly nice girl of a sweet
frank, cheerful disposition. I think she will make
an excellent wife for him. When you receive this
letter you will all be assembled at your hard-named
rural retreat in Germany,* and I heartily hope you
will all enjoy it, and have good health and much

---

* The family party of Horners, Pertz, and Lyells, assembled in the summer
of 1858, and passed some weeks at the small village of Zwingenberg on the
Bergstresse at the foot of the hill of Melebocus between Darmstadt
and Heidelberg.

amusement. Is Melebocus a mountain? What a 1858.
pastoral name, calling up associations of Corydon
and Phyllis?

Humboldt is indeed a wonderful man, but I fear
there is little chance of his living to complete the
remaining part of his work, which is to embrace the
whole of the organic world.

I wish you could have been at our dance,—it was
great fun. With much love to all your party, I am
ever,

Your very affectionate Brother,
C. J. F. BUNBURY

JOURNAL.

July 7th.

In my opinion, Joseph Hooker is now (since
Robert Brown's death) by far the greatest botanist
of this country, if not of Europe. The other
botanists of this country whom I consider as of
the *first* class, are : Sir William Hooker, Bentham,
and Lindley.

In the second class are Henslow, Harvey, Bennett, Thomson, and many more.

July 10th.

Our dear friends Charles and Mary Lyell arrived
in the evening.

July 11th.

The newspapers mention the death of Bonpland,
at the age of eighty-five. When we were at Berlin,

1858. in '55, Humboldt told me that he had lately heard from Bonpland, who was then, although above the age of eighty, planning a voyage up one of the great rivers of Brazil or Paraguay. Very lately, Mary tells me, Humboldt wrote to one of his American friends that Bonpland was still alive and well, and talked of a voyage to Europe. The next mail brought the account of his death. My Uncle Mr. Fox knew Bonpland at Buenos Ayres, after he had been released from his captivity in Paraguay, and said he had then become very *un-European.*

The association of Humboldt and Bonpland was a very fortunate one, they were well suited to be fellow-workers in science, each having the qualities which the other wanted ; for Bonpland was altogether a man of detail and minute accuracy, with little taste or talent (my Uncle said) for large generalizations, and caring for nothing but systematic and descriptive botany.

<hr />

July 13th.

The Lyells left us to return to London. I need not repeat what I have so often said in my journals, of my love and admiration for both of them,— sentiments which I feel increase rather than diminish in proportion as I know them longer. In fact I do not know any man towards whom I feel more of love, esteem, admiration and confidence, than I do towards Charles Lyell ; and his Wife is worthy of him. Lyell's thoughts are at present very much engaged by Darwin's speculations on the great question of species in natural history, 1858. and the opposite views of Agassiz. Darwin has been engaged for nearly 20 years in a work on the general question of species, which is not yet nearly ready for publication ; but at last, as there was some danger of his being forestalled,—for Mr. Wallace, who is employed as a natural history collector in the Eastern Islands, had independently taken up some of the same theories, and sent home a Paper containing his own views,—therefore Lyell and J. Hooker persuaded Darwin to allow one chapter of his work to be published.

This chapter, written fourteen years ago, and containing as it seems the pith and essence of his theory, was brought before the Linnean Society by Lyell and Hooker, with a preface of their own to explain its history ; and they have thus made themselves in a manner, sponsors to it, though neither of them, as I understand, is prepared altogether to adopt Darwin's views.

Darwin has arrived at the conclusion that there is really no such thing as species ; that the great law of nature in the organic world is that of unlimited variation ; and that by the action of this law under the influence of external circumstances, in an indefinite lapse of time, any form of organic life may be derived from any other. In short he believes in " the transmutation of species." But he differs from the Lamarckians in this, as Lyell tells me, that he sees no reason to believe in a regularly *ascending* series of changes, a regular progressive development ; he holds that variation may as often

1858. and as easily take place in the sense of degeneration
as the contrary. Agassiz, on the other hand, as
Lyell tells me, argues strongly for the fixity of
species, and contends that the different forms which
have successively appeared at different times have
been produced not by variation or transmutation
out of those previously existing, but by direct acts of
creative energy. Lyell thinks there is great force
in his reasoning.

Much of Darwin's argument is built upon the
varieties of domesticated animals, such as dogs and
pigeons; but Agassiz meets this by contending that
the numerous domestic races—of the dog for
instance—are not derived from one single specific
stock, but from three or four or more; as has been
already maintained (in the case of the dog and the
horse) by Hamilton Smith. Lyell says that Agassiz
is certainly a zoologist of very high authority and
merit, but that as he has a particularly good eye for
minute differences and distinctions, so his tendency
is rather to rely too much on these, and rather to
multiply species too much than the contrary. His
knowledge of botany also is considerable.

Lyell says that Darwin thinks him (L.) incon-
sistent, in maintaining the doctrine of uniformity
in geology, and at the same time believing in the
creation instead of the transmutation of species.
But he contends that there is no inconsistency,
since he holds the creation of new species to be an
act that is still going on from time to time; not one
that belonged only to former ages of the world.

———

Mary Lyell told me the other day, that she had visited Robert Brown just a week before his death and found him lying in the room which had been Sir Joseph Banks's library, where she had so often seen him before, and where I have often seen and talked with him. He talked quite calmly and cheerfully, recalling the days when he had sate in the same room in company with Banks, Solander and Dryander, and telling her *where* each of them used habitually to sit.

Mary told us rather an amusing anecdote of Mr. Motley, the American historian of the Dutch Republic. When he was a very young man, amusing himself in Europe, he lived very *fast*, and spent so much money, that his Father wrote to him to reprove his extravagance, and to say that he must not indulge in so many luxuries. Motley's reply was to this effect :—"My dear Father, the *necessaries* of life I might perhaps contrive to dispense with, but its *luxuries* I could not possibly do without !"

We talked one day of the diminution of wit :— how there are none now-a-days to supply the place of Sydney Smith, Theodore Hook, or Samuel Rogers. Lyell did not think this was owing to any general cause, such as anything in the state of society, or in public opinion ; but that it was (what we call) casual, just as there was a much greater number of eminent poets in the first quarter of the century than now. I doubt.

1858.   July 24th.

I finished reading Sir Charles Napier's " Romance of William the Conqueror."

_____

July 26th.

We went by railway to Ipswich, and thence in a fly to Mr. Thomas Mills's at Stutton. In the railway carriage we met, first, Mr. Dashwood with his Wife,* a very handsome woman ; and secondly, at the Haughley junction, our agreeable friend Mr. Bowyer, the inspector of workhouse schools, with whom we had some talk both on literary and political matters. He had been reading Froude's History of England, and he remarked that Mr. Froude lays the white-wash very thick on Henry the Eighth, and that some of it sticks.

The party at Stutton besides Mr. and Mrs. Mills and ourselves, were young Mr. Barnardiston, Mr. Mills's nephew, and his pretty young bride, Lady Florence (by birth F. Legge), and Mr. Zincke, whom we have known some years :—a clever man, and a zealous and active champion of education. Mrs. Mills was one of the Barringtons, a very lively, frank, agreeable person.

_____

July 27th.

I went early into Ipswich and served on the grand jury at the assizes. Lord Campbell in his charge, was very complimentary to the magistracy of Suffolk. We had only two very serious cases before us.

* Afterwards Lady William Graham.

A pleasant, cheerful evening at Stutton. I forgot 1858. to mention, in the party there, Mr. Barrington Mills, Mr. Mills's Son by his first Wife, a sensible man.

————

July 28th.

Spent the morning and part of the afternoon quietly and pleasantly at Stutton. It is a very pretty place, situated about 7 miles from Ipswich, on that tongue of land between the estuaries of the Orwell and the Stour which is much the best part of Suffolk. Beautiful woods and lawns sloping down to the bank of the Stour : the broad estuary looking (when the tide is in) like a fine lake, seen in bright peeps between the noble trees. The evergreens at Stutton are remarkably fine : I should think not surpassed by any on the eastern side of England.

A group of Ilexes of extraordinary size and beauty; an uncommonly fine and lofty Cypress, with a trunk measuring 6ft. 10in round at 4ft. from the ground : some of the noblest Cedars and Silver Firs I have ever seen, and a Portugal Laurel forming (from one root) a dense unbroken mass of verdure, 270 feet in circumference. In the morning Mr. Mills showed us a number of really exquisite drawings done by his first wife to illustrate a journal of a tour. Afterwards we strolled about the beautiful grounds and along a pleasant shady walk on the bank of the river ; and all the company, except Lady Florence and myself, went on the water in Barrington Mill's yacht.

1858.   At three o'clock we set off for the Manningtree
station, and thence by railway to London.

Mr. Mills said that he had seen one of the new
reaping machines at work; that it worked very well
where the corn was standing well up, but did not
answer at all where the corn had been in any degree
laid by wind or rain.

————

July 29th.

Harley Street.

After visiting Sir James Clark, we went to the
British Institution to see the exhibition of pictures
by the old masters—a very good one.   What
pleased me most was the " Manfrini Giorgione,"
the famous three portraits, celebrated by Lord
Byron in " Beppo."   I had seen it at Venice nearly
thirty years ago.   It is certainly a most exquisite
picture and deserves all that Lord Byron says of it ;
it appears that he made a mistake about the subject
as Giorgione never had either wife or son.   Mrs.
Jameson says they are a Venetian Lady, a Cavalier
and a Page, whose names are unknown.   The lady,
who is very bewitching indeed, appears to be the
same person as Titian's " Flora," at any rate very
like her.

The " Vierge aux Rochers," by Leon. da Vinci,
a singular and striking picture ; the Infant Christ
very fine ; but the colouring strangely cadaverous.
The landscape is very peculiar.   It almost seems as
if the artist had intended to give an effect of *blue
light* (like that of the " Grotta Azzurra " at Capri,
which however I believe was not known in his time)

and hence the cadaverous colouring. Two Heads, 1858.
apparently original studies by L. da Vinci for his
fresco of "The Last Supper," very grand.

A noble portrait of Charles the Fifth, by Titian.
Several excellent portraits by Sir Joshua Reynolds,
especially a fascinating one of " Nelly O'Brien."

Dined with the Charles Youngs: a pleasant
party; Mrs. Pellew, Fanny Pellew and her betrothed
Louis Mallet, the Douglas Galtons, a young Mr.
and Mrs. Winthrop, from Massachusetts, Mr.
Thomas Baring, and Edward. Nothing particular
to record of the conversation.

—— ——

July 30th.

Shopping with Fanny for two hours in the morning
—then to the British Museum. Looked at part of
the fossil plants, but got no new ideas; this is one of
the most imperfect parts of the natural history
collections there, and particularly ill arranged.
Then looked over a part of the beautiful collection
of minerals, wishing to revive in some degree my
knowledge of them.

Saw also some extremely beautiful exotic lepi-
doptera and beetles. The collection of fossil fishes
is immense, and seems very carefully arranged.

Dined with the Douglas Galtons. The party :—
Sir Alexander and Lady Tulloch (the famous
Colonel Tulloch of the Crimean inquiry), Laura
Bonham Carter, Mr. Lacaita, Douglas Galton's
brother (I do not know his name), and Edward. Sir
A. Tulloch seems a grave, quiet, sensible Scotchman,

1858. not very talkative.   He thinks the war in India will last for years.   Douglas Galton remarked that it was unfortunate the communications in India by railways and electric telegraph had not been made some years sooner.   Sir Alexander said that they would certainly have given us a decisive superiority if they could have been completed *before* the mutiny; but he thought that if these works had been begun some years ago, the mutiny would only have broken out so many years earlier; that the chiefs of the malcontents perceived the superiority which those works would give us, and had thought that "now or never," was their time to break out, before our power was effectually confirmed by those means of communication.

Mr. Lacaita gave us a most spirited and entertaining account of his visit to Eton in company with the Duke of Malakoff; illustrating most amusingly the excessively familiar and unceremonious manners of that soldier.

———

July 31st.

Looked at the Water-colour Exhibition, but slightly.   At my club, I read in the *Athenæum* newspaper an interesting letter from Humboldt, about his old friend Bonpland.   He does not think Bonpland's death yet quite certain.   The last letter he, Humboldt, had from him was written in June, 1857, from Santa Anna in Corrientes, and he then wrote in good spirits.   Humboldt had very lately heard of him from Dr. Lallemant, who saw him at the same Santa Anna, where he has been living for some years past, cultivating Orange trees on a large 1858. scale. His habits of life were extremely abstemious.

We returned to Mildenhall by the 5 p.m. train.

August 6th.

Sir John Walsham, who came to luncheon with us after visiting the Board of Guardians, brought us the first news of the electric telegraph cable having been successfully laid, and the sub-marine communication actually opened between Ireland and Newfoundland. This is really a wonderful achievement, Sir John mentioned that he had lately seen a reaping machine at work at one end of a field, and a thrashing machine at the other end of the same : so that the corn was no sooner cut than it was delivered over to the thrashing machine.

This is glorious weather for the harvest, which is going on rapidly. We saw the harvest beginning in some places even as we went to Ipswich.

August 9th.

Drove over to Barton to see my Father, and returned in the evening. An extremely hot day. In the arboretum at Barton, the large tree of Abies Cephalonica, originally raised from one of the seeds sent home by Sir Charles Napier, has several cones on it for the first time : it is I believe, the second tree of the kind that has borne fruit in England ; Mr. Long's,* in Surrey, being the first. Abies Smithiana is also bearing cones this year, for the first time at Barton.

* At Hampton Lodge in Surrey.

1858.   August 10th.

Our friends Mr. and Mrs. and Miss Rickards, and Mr. and Mrs. Bowyer arrived.

—————

August 11th.

Spent the morning very agreeably in showing my collections to Mr. Mrs. and Miss Rickards. They all three have a hearty love for natural history, with very considerable knowledge of it, and great habits of observation.

—————

August 12th.

To Ely, a party of seven : spent a pleasant day there.   Admired the cathedral both inside and out, especially the exquisite carved screen behind the altar, and the magnificent coloured glass of the great east window.   Attended afternoon service in the cathedral ; saw Dean Peacock and his agreeable wife,* and ended with a visit to the Bishop,† who was most courteous, and showed us his garden and as many of his pictures and multifarious works of art, as the lowering darkness (for a storm was rapidly coming on) would allow.   There are noble trees in his garden : the finest American Plane (Platanus occidentalis) I think that I ever saw ; a magnificent Walnut tree, and a very fine Catalpa, loaded with blossoms.   This was an intensely hot day, but thunder and rain came on in the evening.

* Sister of Bishop Selwyn, afterwards married Mr. Thompson, Master of Trinity College Cambridge.

† Bishop Turton.

The Rickardses left us. Mr. Rickards is a remarkable man : of clear and powerful intellect, with a range of reading and knowledge, which he brings into play in conversation with a very agreeable frankness and simplicity. He seems to be a profound classical scholar, extensively conversant also in natural history, in English antiquities, in architecture, in the literature of our best times as well as the more modern, and to some extent even in the theory of war ; in fact, I hardly know a subject of enlightened conversation in which he is not qualified to take a part. Add to this, he is an excellent man, of an engaging simplicity of manners and character.

LETTERS.

Mildenhall,
October 29th, 1858.

My Dear Emily,

We returned yesterday evening from Hardwick, where we had spent three days very pleasantly indeed ; the little Lady most kind, genial, and agreeable. There were staying in the house three young friends of hers, with whom we were quite charmed ;—two sons of the noted Colonel Birch (the military Secretary at Calcutta), whose wife you know is a very old and dear friend of Lady Cullum's, —together with the young wife of one of them. Whatever Colonel Birch may be, his two sons are singularly pleasing young men,—full of spirit and

1858. intelligence, with remarkably agreeable manners;
judging from their conversation, thorough gentlemen
in feeling as well as in manners: and moreover,
very handsome. The elder brother, Captain Birch,
was in Lucknow throughout the siege, as Aide de
Camp, first to Sir Henry Lawrence and then to Sir
John Inglis; the younger was in Havelock's army,
and was present at the relief of Lucknow. The
meeting of the two brothers under such circumstances must have been most interesting. They
seem affectionately attached to each other. The
elder is newly-married to a very young and delicately
pretty wife: quite a Madonna face. We had some
very pleasant strolls with Lady Cullum and these
young people. Tuesday and Wednesday were two
of the most exquisite days that I ever saw or felt,
and never did I see Hardwick look more lovely;
indeed I think there is something in these beautiful
autumn days, with their soft brightness and the
various rich tints of the trees and the delicate
colouring of the distance that sets off our English
landscapes to more advantage than any other
season. Yesterday again it was a continued *soak*
from morning to night, and to-day a cold blustering
north wind.

I think I have not written to you since young
George Blake was here: we were very much pleased
with him: but he could stay only one night. He
and the younger of the two Birches knew each other
well when in Havelock's corps. We saw him
again (Blake I mean) together with his Uncle
Patrick at a large dinner party that Lady Cullum

gave on Wednesday; whereat were also Colonel and Lady North, Lady William Hervey, and the second Miss Seymour, and some others.

I did receive your letter of the 14th, and was much obliged to you for it. As you do not mention my Father in your letter to Fanny, I trust he is in pretty good health. I am sorry you have had so much bad weather. — Our visitors left us last Wednesday week: we were very sorry to part with Augusta and her children. She is one of my especial favourites, and I think Fanny was quite in love with little Freddy, who is indeed a very fine little fellow.

There was very little business at the Quarter Sessions: only eight prisoners for trial. The inquiry concerning the Master of our Workhouse, which I mentioned in a letter to my Father, ended just as I expected,—very unsatisfactorily; the decision of the Poor Law Board is equivalent to a verdict of *not* proven.

Pray give my love to my Father, and believe me,

Ever your affectionate Stepson.

C. J. F. BUNBURY.

P.S.—Charles Lyell is returned safe and sound from Sicily.

Mildenhall,

October 31st, 1858.

My Dear Katharine,

I thank you very much for your kind and agreeable letter, which I received at Hardwick. I do indeed most heartily rejoice to hear that your dear children are really convalescent, and that you

1858.   are released from those terrible fears and anxieties
you have so long been enduring on their account.
I trust you will have before long the happiness of
seeing the dear little fellows looking as well as ever.
Your stay at the Crystal Palace hotel must have
been very pleasant, with the opportunities of seeing
that beautiful and interesting place so quietly and in
such leisure. Is it true that they are talking of another
grand exhibition in '61 ?   I hope the Sydenham
Palace will not meet with such a catastrophe as has
befallen the New York one.

Yes, I will certainly bring the parcel of Ferns
which I have laid by for you, containing specimens
of Doodia candata, Adiantum pubescens and Pteris
(Pellaea) hastata, from our greenhouse, together with
a few Brazilian ones which have turned up since you
were here. We have now about twenty-four
species of Ferns flourishing in our greenhouse,
besides a few that are in less thriving condition ;
and out of doors we have my favourite Polypodium
Dryopteris growing nicely, as well as most of the
common English ones.   Pteris serrulata seeds
itself to an extraordinary extent in the greenhouse
and its little delicate germinating fronds are very
pretty.

This fine warm season has given us some really
tolerable out-door grapes from our wall.   We have
had no frost yet, and the tea-scented China Rose is
still in blossom, as well as the Scarlet Geraniums, a
variety of Verbenas, and some other things.   We
have planted a fine young Deodar from Barton, in
our Paddock.

It was a great comfort to hear of dear Charles 1858. Lyell's safe return. I look forward with very great pleasure to our all meeting in London.

With much love to Harry and your children, and all the family,

I am ever,

Your most affectionate Brother,

C. J. F. BUNBURY.

———

18, German Street, Brighton,
December 10th, 1858.

My Dear Emily,

I was very much obliged to you for your letter, but very sorry indeed to hear such an uncomfortable account of my dear Father; I had not at all understood from Edward's account that he had been so seriously ill on the journey; but as you say, Mr. Perry thought well of him at his last visit, I trust he has gone on satisfactorily, and has by this time got over the effects of the journey. I shall be anxious till I hear again of him.

The weather lately has been sadly unfavourable for delicate people; this raw, foggy cold succeeding so suddenly to the mild and bright weather of last week, is very trying; and I am afraid you and my Father must find it very gloomy and uncomfortable in London, where the fog seems to be very thick.

We shall remain here (God Willing, as our forefathers used to say) till next Thursday, and then go to Oaklands, where we propose to remain a week, then to Folkestone for another week, before

1858 returning to London. I keep wonderfully well through all this bad weather.

We have made acquaintance with Lady Louisa Kerr,* who is living here. We spent a couple of hours with her last Monday: she was very friendly; said she well remembered being at Mildenhall† with her Father and Mother, and remembered me as a little boy, and asked much after my Father. She is a good talker, a very eager politician, and has a passion for the fine arts. She is very fond of Mr. Buckle, and proposes to invite us to meet him. She says he is a *shy* man! I should have supposed his notions of merit and modesty were much the same as Sydney Smith's.

I hope you have seen the last *Punch;* " Pam " recommending to Britannia the new Parisian fashion of handcuffs.

What a curious and what a disagreeable business was the theft of the Ionian dispatches. I suppose you will not remain very long in town.

Pray give my love to my Father, and believe me,

Your affectionate Stepson,

C. J. F. BUNBURY.

———

6, Pleydell Gardens. Folkestone,
December 24th, 1858.

My Dear Katharine,

A merry and a happy Christmas to you and yours, with all possible good wishes, and very many thanks for your kind and agreeable letter,

---

* Daughter of Lord Mark Kerr and Lady Antrim.

† In 1816.

which I received yesterday. I do sincerely admire 1858.
your industry and zeal in working at your botany in
the midst of so many distractions and household
cares, and truly glad I am when I can help you in
any way. I shall be delighted to see your arrangement of Ferns, and to explain more fully than I can
do by letter, why I have deviated from Sir Wm.
Hooker's arrangement in the instances you mention.
I am delighted to hear that you are also turning
your attention to the Mosses; they are charming
little plants, great favourites of mine, and I hope to
resume the examination and arrangement of those I
have when I am settled again in my den at home.
You must have some very interesting ones. I quite
forget what was the list of genera that I gave you;
I suppose taken from Bridel; but as soon as we go
home, I will copy out for you the arrangement
given in Berkley's "Cryptogamic Botany" (which
unluckily I have not with me here), with such
explanations as I can give.

I have not, as you may suppose, had the opportunity of doing much practically in botany; but in a
walk with the charming fairy Sarah (she is more
lovely almost than ever) at Oaklands, I showed her
some Mosses in fructification, and perhaps laid the
foundation of a taste for them. She is certainly a
singularly attractive little girl, and of a most sweet
disposition.

We spent six very pleasant days at Oaklands, and
nothing could exceed the warmth and kindness of
our friends there; but poor Minnie was far from
well, very much shaken and pulled down by a severe

1858. feverish cold.   Oaklands agreed very well with me.
We left it the day before yesterday, and slept that
night at St Leonard's, where we saw my cousin
John (who is deaf and dumb) with his nice wife and
pretty children ; and yesterday we came on hither
in a furious storm of wind and rain, which lasted
till past the middle of the night.   To-day has been
fine and comparatively calm.

I cannot say I have fallen in love with Folkestone
but I am glad to be here for a few days for the sake
of seeing Henry and Cecilia, and their dear children.
Henry has been out of health and seems out of
spirits, poor fellow ; Cecilia well ;   the two elder
children very pretty (the boy especially), well grown,
very merry and full of fun, and very much at their
ease with us ; the youngest (who was a very young
baby when I saw him before) much grown, and a
fine little fellow.   The William Nicholsons are in a
house very near us, and Fanny has seen them.   I,
not yet.

I am very glad to hear so good an account of all
your party, and that you are all so merry.   I am
not yet sure whether we shall be able to come to
you on the 29th, though I should like to see your
dear children's merriment.

I am very glad indeed to hear of any addition to
Joseph Hooker's comforts, and heartily agree with
you in wishing that he had a great deal more.
Really when one sees men not worth his little finger,
rolling in wealth, one feels rather dissatisfied with
the world.

I see in the newspapers that the botanical room

at the British Museum, which poor Mr. Brown was 1858.
so long engaged in arranging, has been opened to
the public. I remember that the last time I saw
him at the Museum he showed to Fanny and me
the collections in that room ; they were very
interesting.

With much love to all your party,

I am ever your very affectionate Brother,

C. J. F. Bunbury.

JOURNAL.

January 1st.

A family dinner of the Horners and Lyells.   1859.

Charles Lyell told us of the good service that
Lord Stanley had done by preserving the Botanical
Museum (formed by Dr. Royle) at the East India
House, which it had been intended to break up. It
is a very valuable collection, illustrating particularly
the *economic* botany of India, and its industrial
resources connected with the vegetable kingdom.
Lord Stanley seems to be doing admirably well in
his new Indian department.

January 4th.

We dined with the Horners ; met Joseph Hooker
and his Wife. He had brought and showed us
some very curious specimens — fossil cones of
Banksia, collected in New South Wales, not far
from Sydney, in a bed which is covered by a great
thickness of basalt ; and he had brought also a cone
of the recent Banksia ericifolia, to show the wonder-

1859. fully close agreement between them; the fossil showing the thick shell-like follicles (with even the lines of suture distinguishable) projecting from the mass of withered remains of flowers, the unexpanded flowers packed together at the top of the cone, and in one instance a follicle open with the winged seed visible. It is rare indeed to see vegetable fossils so clearly identifiable with recent forms. The age of these curious remains is uncertain, but believed to be older than the bone caves.

Hooker talked a good deal of the much discussed separation of the British Museum collections—the separation of the natural history collections from the antiquities. It seems to be now almost certain that this separation will be made, and Hooker said that for his own part, though at first much averse to the scheme of establishing the natural history collections at South Kensington, he had after much discussion become convinced that it would be on the whole the most eligible site. He said that an admirable plan had been drawn up by Huxley and Bentham for the arrangement of the natural history collections, in case of their being established in a new building. It seems the favourite plan is to transfer the botanical department of the Museum to Kew; but Hooker very justly says that in *that* case, a select typical herbarium ought to be kept in London. He says there are whole rooms in the underground story of the British Museum, entirely filled with packages of dried plants which have remained for scores of years unexamined and unopened.

Dined at the Geological Club; present,—John Phillips (president), Lord Enniskillen, Mr. Horner, John Moore, William Hamilton, Prestwich, Prof. Miller of Cambridge, Mr. Mylne, Dr. Bigsby, and one or two others. Pleasant talk. Lord Enniskillen is a very good fellow and a thorough Irishman. He made us all laugh by saying of the famous fossilist Count Münster, that he was very stingy and unfair, always *making exchanges and giving nothing in return.*

Hamilton said that when he was in Asia Minor, Europeans could travel with safety through most parts of the country; *now* the whole country swarms with *Greek* robbers, so that no one can travel without a strong escort, and these bandits infest even the immediate neighbourhood of Smyrna, where he was able to ramble alone with perfect safety.

Evening meeting of the Geological Society.

There was a rather interesting discussion between C. Lyell, W. Hamilton, and Prof. Miller, on the question whether serpentine is always a metamorphic rock or sometimes intrusive ; the author of the Paper had maintained the first doctrine, Lyell and Hamilton held the other.

January 6th.

In talking of the sensation created by the French Emperor's words to Count Hübner on New Year's Day, Lyell said that Mr. Pulsky has for months past maintained that whatever other demonstrations the Emperor might make, however he might appear

1859. sometimes to threaten England, his real object is
to drive the Austrians out of Italy. By this move,
besides imitating his Uncle and gratifying the army
he might likely enough be able to make an arrangement with the King of Sardinia, by which Savoy
might be transferred to France, in return for assistance in conquering Italy.

———

January 7th.

We went to Clapham and saw my Uncle Sir
William Napier. He has rallied wonderfully from
the very brink of the grave, but I fear there is
no prospect of his ever really recovering so far as to
be free from suffering or to have any enjoyment
of life. He is utterly crippled and reduced to skin
and bone, but his head looks even grander and
more eagle-like than before his illness; his voice is
strong, he talks incessantly and with wonderful
fluency, clearness, and energy. I fear he will have
much suffering yet before his death; a sad price
to pay for a strong constitution.

In talking in the evening of English poets, Lyell
praised Cowper more highly than I had been
disposed to do, but remarked that his fame must
rest chiefly on his minor poems, the large ones being
very prosaic.

He agreed with me in thinking that "Gray's
Elegy" would last as long as the language,
and said that he had found it more universally
known in America than any other English
composition—known to every school-boy and school-

girl. Daniel Webster had it read to him the day 1859. before he died. This is an interesting counterpart to the anecdote of Wolfe.

January 8th.

A very pleasant dinner party, and a large party in the evening. Lady Bell spoke of Sir William Napier as being, in 1812, when she first saw him, the most glorious specimen of human beauty she ever beheld. She first saw him at the play when she and her husband were in the same box with him, and his brother George and *his* bride, and their Mother, Lady Sarah ; and it was beautiful to see the devotion of the two young men to their Mother.

Read in Agassiz's work on Lake Superior, the portion relating to botanical geography :—ably written and containing good information. What is odd, he speaks of certain butterflies, such as Papilio cardui, Atalanta, and Antiopa, as having been *introduced* into America from Europe *together* with European weeds. It seems difficult to conceive how this could have happened, as those weeds must surely have been introduced in the state of seeds.

January 10th.

Went with Charles Lyell to see the Botanic Garden in the Regents Park. The great American Aloe (Agave), which flowered here last summer, is still a conspicuous object ; its flower stem (24 feet high) a beautiful chandelier-like panicle of flowers, though completely withered, still standing up entire

1859. and unbroken, and making a fine appearance. In the large conservatory I saw nothing altogether new. Araucaria excelsa, nearly up to the roof, so that its growth must soon be checked : Araucaria Bidwillii, very flourishing : Grevillea robusta, a beautiful tree with drooping branches and delicate fern-like foliage (quite unlike the hard stiff character of the Proteas and Banksias) also nearly reaching the roof.

We left the dear Lyells, with whom we had spent ten very pleasant days, and removed to Mr. Horner's.

January 11th.

At the Athenæum, read in the papers the good news of the release of Poerio and other unfortunate political prisoners at Naples from the horrible dungeons in which they have so long been tormented. It is a comfort to hear of this, though one cannot give much credit to the King of Naples for his motives.

The rumours of approaching war on the Continent are gaining strength and consistency : it is evidently wished for at Turin, and thought probable at Paris ; the French funds are going down, and the Austrians are strengthening their armies in Italy.

Began to read Bulwer's new novel—" What will he do with it ?"

January 16th.

We walked in the Zoological Gardens : looked at the Wolverene (or Glutton) a new acquisition, but

he was sleepy and would not show more than his 1859.
nose and fore paws ; he looks like a little bear.
The Fennec fox, another novelty from Egypt
(Nubia ?) is a beautiful little creature to which
the figures I have seen in books do no justice at all.
The " clouded tiger " from Assam is another animal
of remarkable beauty ; this indeed I had seen
before. The elk, on the contrary, remarkably
clumsy and uncouth, as ugly as a quadruped can
well be, a huge creature as high as a large horse ;
he is at present without horns. Afterwards I spent
sometime with Katharine, looking over Mosses
with her. She showed me a set of specimens of
Mosses given more than 40 years ago by Signor
Raddi to Mr. Lyell, named in Raddi's handwriting,
but without localities, but they must evidently have
been collected by Raddi, in Brazil, and are almost
without exception identical with the species that I
found in Brazil. She showed me also a large
quantity of Mosses from Simla, of her own
collecting, many of them fine things ; so distinct
both from the European and from the South
American Mosses that I recognised hardly any except Hypnum proliferum.

In talking of novels with Mr. Horner this evening,
he told me that not only was Sir J. Mackintosh
a most assiduous reader of novels, but Sir S. Romilly
was so likewise, in spite of his overwhelming mass of
occupations in law and politics. Mr. Horner
mentioned that one day in a company where he and
Romilly were both present (Romilly at the time in
the height of his reputation and employment), *he*

1859. (Mr. Horner) and another man were discussing one
of the Waverley novels—the then last new one ;
they agreed that some one character in it might
have been omitted or made less prominent ; Romilly
who had been talking of something quite different
suddenly turned round to them and said, " I would
not omit one word."

Mr. Horner's birthday.   Went to the Linnean
Society. Called on Mr. Boxall.

A family dinner party at Queen's Road West,
and a sociable evening. News of Cecilia's confine-
ment and the birth of a third little boy.*

Fanny and I visited Edward at his lodgings.   He
told us he had just been elected a member of a new
club, " The Alpine," and showed us a sheet of an
extremely beautiful Swiss map of the Alps. We drank
tea with Mrs. Shirreff. Met the Richard Napiers and
William Greys.   Mrs. Grey is very agreeable, and
there are few people in whose society I delight more
than in that of the Richard Napiers.

We called on Mrs. Maurice, an old friend of Fanny's,
Sister of Julius Hare, and second wife of the celebrated "broad church" divine, Frederick Maurice.

Dined at the Geological Club.   At the evening
meeting of the Society was read a paper by Mr.
Rosales on the gold field of Ballarat in Victoria ;

* Captain William St. Pierre Bunbury, R.A.

not very interesting in itself, but it gave occasion to 1859. a good discussion, branching out into a variety of topics in which Murchison and Lyell took the most prominent part. Among other things the author of the paper alluded to the great streams of basalt or lava overlying the auriferous drift, and to the occurrence of vegetable remains, and in particular of a cone of the " she-Oak " (Casuarina) under the basalt. This gave me occasion to speak of the vegetation, recent and fossil, of Australia, and especially of the Banksia cones now in Joseph Hooker's possession. By the way, Murchison says that those cones were found in the Colony of Victoria, not in New South Wales. Murchison was very copious and instructive on the subject of the gold deposit, and of the occurrence of gold in general.

January 20th.

Went to the South Kensington Museum, principally to see the Sheepshanks gallery of pictures : a very interesting and agreeable collection.

The set of Landseer's is particularly good, many of his best things are here, especially " The Shepherd's Chief Mourner," and " Jack in Office."

Attended the evening meeting of the Linnean Society.

I had a good talk with Bentham, especially on the subject of species and varieties, on which no one is more qualified to give an opinion than himself. He said that he has not been able to discover any absolute *test* for species : that the

4 K 2

1859. question—What is a species ?—must be determined
in each individual case by a minute and careful
examination of circumstantial evidence, and is often
very difficult. The test of cultivation, he said,
requires much caution in its application, and he
believes that many of the recorded instances are
fallacious.   One very common source of such
fallacy is, that the special plants under cultivation
die, more common and hardy ones, of which the
seeds were in the soil, spring up in their place
and a change is believed to have taken place.   In
this way, he believes, is to be explained the supposed
proof of the identity of Trifolium hybridum with
Trifolium repens.   He is not entirely satisfied as
to the identity of the Cowslip and Primrose, which
is believed to have been established by cultivation.
He remarked that Linnaeus had a peculiar and sure
tact in discerning natural species, though he fell into
some errors by believing the influence of *hybridization*
to extend much further than it really does.   It is by
no means yet a settled point whether any of the
varieties found in a wild state are to be considered
as hybrids.

Bentham spoke of the numerous forms of Oaks
intermediate between Quercus pedunculata and
sessiliflora, which occur in some of the hilly
districts of England, he does not believe as Lindley
does, that these are natural hybrids between two
distinct species, but considers them as simply
varieties tending to connect the two extreme
forms of one variable species.   These intermediate
forms are rare in the plains ; Quercus pubescens

he thinks, is another variety, or *race* (sub-permanent 1859. variety) of Quercus Robur, confined to the Mediterranean countries.

Again, the Cork tree in Bentham's opinion, is only a variety of Quercus Ilex, as indeed, Sir J. E. Smith long since suspected (See Ree's Cyclopædia) and the Quercus Ballota and Quercus Gramnutia are likewise forms of the same excessively variable species.

The Luccombe Oak is really a hybrid between the Cork tree and Cerris, but Bentham does not believe it is ever found wild.

January 21st.

Went with Fanny and Joanna to the British Museum, where we examined the Greek vases : Susan joined us, and got leave from Mr. Birch for us to see the Halicarnassian marbles, which are not yet opened to the public. There are very beautiful friezes in high relief of combats of Greeks and Amazons, wonderfully fine in form and action, and two grand colossal figures supposed of (?) Mausolus and Artemisia : the female figure has lost its head : the head of the male remains, and is evidently a portrait.

January 22nd.

To Clapham, to see my Uncle and Aunt. Found Sir William in much the same state as on the 7th : at first indeed rather languid and depressed, but he rallied as he went on talking, and latterly talked on military matters with wonderful animation and fluency. He expressed a strong opinion (which he

1859. said he had held for months) that the Emperor
Napoleon will go to war with Austria, and join with
Sardinia to drive the Austrians out of Italy. The
formidable army of France will, he said, compel the
Emperor to go to war, and the invasion of Italy will
be attended with much less difficulty and risk,
and offer much more temptation than an invasion
of England. He expatiated on all the military
bearings of the question with a really astonishing
command of argument and force of expression.
Then he went on to talk with equal fluency and
energy of the new Armstrong gun, which he thinks
will make a complete revolution in the methods and
prospects of war: principally by its effect in naval
engagements and against sea-coast towns: on land
he thinks its principal effect will be to render battles
more distant and more indecisive. Sir William's
power of discourse (for it is not *conversation*) is
wonderful in any man, but especially in one who has
a mortal disease, and who was so lately on the very
brink of the grave.

_____

January 23rd.

We went to Twickenham, and spent a very
agreeable afternoon and evening with our dear and
valuable friends the Richard Napiers.

_____

January 24th.

To Sandhurst, to the William Napiers: William
being Superintendent of Studies in the Military
College.

_____

January 25th.

At Sandhurst. In the morning I walked with

1859.

William about the grounds and through the College, and he showed me in particular the Model room where I was much interested by the large and well executed models of fortified places, showing all the distinctive characters of Vauban's and Cormontaign's systems of fortification with the siege operations. These models are very instructive, and particularly so to me, as William was able to explain everything to me. Thus I gained a much clearer idea of the different parts of a fortification and of the operations of a siege—of the outworks, glacis, covered way, places d'armes, trenches, parallels, enfilading and breaching batteries, sap, &c., &c., than one ever can from descriptions or plans on paper. There is also an instructive model on a large scale of Sebastopol and the country round it, showing all the works thrown up by the besieged and by the besiegers, the English and French trenches, the ravines, the fields of battle of Inkermann, Balaklava, and the Ichernaia, and in fact every particular connected with the siege. William pointed out to me how very irregular were the works, both of the allies and of the Russians, owing to the nature and form of the ground.

In the afternoon I walked with William to the new Wellington College, between two and three miles from Sandhurst; saw the head master, Mr. Benson* (lately second master of Rugby) who looks a very shrewd, acute, and resolute man; and saw the boys start off in high glee on a grand paper hunt

* Afterwards Archbishop of Canterbury.

1859. across the country, after the manner described in
" Tom Brown's School Days."

The wild open heathy country about Sandhurst
(on the Bagshot Sands) is very agreeable, appears
very healthy, and must be beautiful when the Heath
is in blossom.

January 26th.

Emily took us in their light open carriage to
Bramshill Park, Sir W. Cope's, where the hounds
met.   The "meet" properly speaking, was over
when we arrived, but the search for the fox was
begun, and as we stood on a bold narrow tongue of
land commanding an extensive view, and looking
across to another plateau covered with fine trees, it
was pretty to see the hounds and the red-coated
horsemen dipping down into the valley, reappearing
among the Fern on the opposite slope, and
scattering here and there as they ascended it to the
wood.   But what pleased me even more than the
beautiful scenery, was meeting Kingsley, who is the
rector of this parish, and to whom I had taken a
great fancy when I met him at Ickworth in '57: I
am very fond of his writings.   We walked back,
Fanny, Emily Napier and I, with Kingsley and Sir
W. Cope to whom he had introduced us, to Bramshill house, and Sir W. showed us the whole of it.
It is a very large and very noble old house, built in
James the First's time; indeed it is said to have
been built for Prince Henry, the eldest son of
James, but not finished when he died.   It was begun
in 1612.   The rooms are very fine.

The park is beautiful with a fine variety of ground
and noble old trees, especially Scotch and silver
Firs : indeed the Scotch Firs are the grandest and
most picturesque I have ever seen, and Kingsley
(who has noticed them in his " Winter Garden ")
says they are the finest in England. The tradition
in the neighbourhood he tells me, is that they were
introduced by James the First from Scotland, at
the same time that he began building Bramshill.
I walked back with Kingsley to his Parsonage at
Eversley, where we had luncheon with him and his
family, the ladies having come thither from
Bramshill in the carriage. He pointed out to me
the relative positions (topographically) of the Upper
Middle and Lower Bagshot Sands, and of the
London Clay, and the peculiar form of ground
characteristic of the Bagshot Sands, — how they
form far-stretching and very flat table-lands, often
very narrow and running out into long tongues
and promontories, with steep sides. He seems
thoroughly conversant with both the geology and
botany of his district, and talked of them in a very
interesting style. He remarked that this wild
heathy country produces many plants of a sub-
alpine character, and he instanced in particular
Lycopodium Selago. Lycopodium clavatum and
inundatum also grow, he said, on these heaths:
and together with the ordinary bog plants, the
Narthecium in great profusion, but no Pinguicula.
He spoke with enthusiasm and in a most interesting
manner of botanical rambles on the Welsh mountains, and of the pleasure of seeing in succession

1859. as one ascends the plants characteristic of different elevations. We formed schemes of future botanical walks in company in his neighbourhood :—whether they will ever be executed, *Dio sa.*—Kingsley said that he thought himself a good judge of the weather in his own district, but at the distance of even forty or fifty miles from his home he found himself completely at a loss in that respect ; so local are the conditions of the weather in this country, depending on form of ground, position relatively to sea and to high grounds and many other circumstances. He told me that it has been found in Devonshire, that the Cork tree bears exposure to the sea air and westerly gales better than any other tree except the Pinaster, better than the Sycamore.

Eversley rectory is small, but a pretty spot with three magnificent Scotch Firs near the house, apparently coeval with those at Bramshill. I was very glad to see Kingsley in his home, and to become acquainted with his wife. He showed me in his library a book of some interest : a copy of an old edition of St. Augustin, given to him by Carlyle, and to Carlyle by John Sterling. He spoke with great admiration of the character and writings of Robertson of Brighton.

January 27th.

We returned to London.

A pleasant little party at Katharine's. Much talk about the " Burns Festival," which had taken place in our absence.

Took dear Kàtharine to the Linnean Society, and showed her the new rooms. Kippist was very attentive, showed us part of the Linnean collections, and of Sir James Smith's correspondence.

January 29th.

We returned home to Mildenhall: found all well and in nice order.

February 3rd.

We went to Barton ; found my Father really very well, considering his age.

February 4th.

My fiftieth birthday.

Lady Bunbury told us that once when she was ten years old, she was taken by her aunt Lady Louisa to see an old lady of 110, a Miss Alexander ; and before they came away the old lady took her by the hand, and said " Now remember, my dear, " you will one day be glad to remember—that you " have yourself seen a person who was at the siege " of Derry."

I have often heard my Step-mother tell this, but this time I will note it down, lest my memory should hereafter fail me as to the particulars.

The siege of Derry was in 1689 ; Lady Bunbury was born in 1783 ; the old Irish lady, being a hundred years older, was therefore born in 1683, and must have been about six years old at the time of the siege.

1859

February 6th.

My Father talked of the time when he was
quartered at Shorncliff in 1803-4, when Napoleon's
invasion was expected; when the officers used
daily to look from the cliffs with their telescopes
towards Boulogne, in expectation of seeing the
French flotilla putting to sea. He thought that
owing to the different rate of drifting of the vessels
the different degrees in which they would have
been influenced by the tides and currents,—the
divisions of the French armament would not all
have reached the same part of the coast; some must
have gone as far west as Pevensey. There was one
choice division of 4000 picked grenadiers, under
Lannes, which was to have been embarked in
row-boats, and would probably have come directly
across to Folkestone or Sandgate; here they
would have been met by Sir J. Moore, who had
about an equal number of the best troops in the
British service; and there would have been such
a fight as has not often been seen. My Father
said that Moore's plan was to attack the enemy
in the water, to charge them while in the very act
of getting out of their boats. My Father is
strongly of opinion that Napoleon the Third does
mean to go to war with Austria this year. He
cannot believe that the Emperor would go on making
such great and costly preparations at Toulon and
elsewhere, without a serious intention of war. He
thinks it likely enough that the combined French
and Sardinian armies may drive the Austrians out

of the Milanese, but that they will be brought to a 1859. stand on the Mincio.

LETTERS.

Barton,
February 6th, 1859.

My Dear Mary,

Very many thanks for your kind letter and good wishes as well as for your pretty and useful present. Indeed I am very fortunate in having such good and kind and true friends as you and Charles Lyell and all my Sisters-in-law.

I feel in very tolerable condition at the conclusion of my *half hundred*, but I certainly cannot fancy myself five-and-thirty, though you very well may in your own case. Lady Bunbury gave me for a birthday present a copy of the same beautiful illustrated edition of "Childe Harold" that you have; and a fine young Wellingtonia was planted on Friday to commemorate the same grand occasion. The analogy between Wellington and me is not perhaps very striking, but that does not much matter.

The pretty Rhododendron Dairicum (or Davuri-cum) of which you sent some flowers in your letters from Sydenham, is in beautiful blossom in the conservatory here. It is from *Daouria*, in the most eastern part of Siberia.

I was much entertained by your account of the opening of Parliament, but I sincerely hope (rather than expect) that you may not have caught any cold or rheumatism from it; it was a great risk.

1859.   I should not have thought the risk compensated
by the pleasure ; but my appetite for pageantry
is very moderate, and I am much afraid of cold.

I do not think the general tone of the speeches on
the first night of the Session very satisfactory as
regarded Italy, and Sardinia in particular ; and
I have some fear that if there is war we shall be
dragged into it on the wrong side.   Our close
connection with Russia makes the risk greater.

I have had a satisfactory and pleasant visit here,
and good talk with my Father, who appears pretty
well, only more deaf than he was ; but I shall be
very glad to find myself at home again, and *very*
glad to see dear Katharine again.   I shall set to
work in earnest as soon as I can at the Indian fossil
plants.   Do not forget that you are *positively* to
come to us this summer and explore the Burwell
pits.

Much love to Charles Lyell,

Ever your very affectionate Brother,

C.. J. F. BUNBURY.

Barton.

Sunday, Feb. 6th, 1859.

My Dear Mrs. Horner,

I thank you very heartily for your kind
note, and for your and Mr. Horner's kind good
wishes on my fiftieth birthday.   Your gifts were
very acceptable, and I value them much, but above
all I value, and shall always value, the assurances
which I so constantly receive, practically as well as
in words, of your good will and kind feeling to me.

I am extremely obliged to Mr. Horner for his map 1859.
of Teneriffe, which I am particularly glad to have.

My Father is tolerably well and cheerful ; and
yesterday, which was a beautiful bright day, he
walked out with me in the pleasure ground and did
not appear particularly weak. Lady Bunbury and
he are both much interested by the Cornwallis
correspondence, especially the part relating to
Ireland in '98. I have begun reading Ellis's
"Madagascar." I was very sorry to leave dear
Katharine, but it could not be helped, as we had
made the arrangement some time before, and it was
all settled that we should be at Barton at this time.
I shall be very sorry if she does not gain health and
strength from the air of Mildenhall, and surprised
too ; and I trust she will stay long enough to give
the air a full and fair chance. We return to-morrow,
and I shall be very glad to see her again.

The conservatory here is looking very pretty with
beautiful Ferns and several flowers, and out of doors
there are Violets in abundance ; but at Hardwick
Lady Cullum showed us beautiful roses from her
greenhouse.

Pray give my love to Mr. Horner.

Ever your affectionate Son-in-law,

C. J. F. BUNBURY.

JOURNAL.

February 7th.

We returned home and found dear Katharine,
and her two youngest children.

1859.   I began reading Atkinson's "Siberia," lent by my
Father.

_____

February 8th.

Began to unpack the fossil plants from India
(from Nagpore) which I have undertaken to
describe for the Geological Society, and to examine
and label some fossil plants from New South Wales
sent me from the geological survey.

_____

LETTERS.

Mildenhall,
February 12th. 1859.

My Dear Susan,

I thank you very much for your agreeable
letter and kind wishes on my birthday. I do most
sincerely feel myself peculiarly fortunate in having
so many kind and true friends, on whose steady
affection I can so confidently rely.

The weather has been wonderfully mild almost
ever since we returned home, and especially since
we came back from Barton; delightful for exercise,
and we have been out several times to the plantations
with Katharine and the children. It is a great
pleasure to have dear Katharine here with us, and
the little ones are very engaging.

What do you think of "Napoleon III. et l'Italie?"
that remarkable state paper, as my Father calls it,
for there can surely be no doubt that, though not
bearing the imperial name, it is really a manifesto
of the imperial policy. It is exceedingly able,

ingenious and forcible. I have as yet, indeed, only 1859. read the translation of it in the *Times*, but shall try to get the original. The *Times* brags mightily about the storm of war having blown over, but I still think there will be war.

I was not satisfied, and I daresay you were not, with the tone as regarded Italy, of any of the speeches in Parliament on the first day of the session, except Lord John's. Even Lord Brougham showed an inclination to Austrianize.

I shall be very glad indeed to see you here, and I hope we shall have a longer visit from you after Manchester. I hope the Superannuation question may be settled in such a manner that Mr. Horner* may be able to retire with comfort this year, as I am sure he will with honour.

Much love to all your home party. I will write to Joanna very soon.

Ever your very affectionate Brother,
C. J. F. BUNBURY.

Mildenhall,
March 5th, 1859.

My Dear Katharine,

Very many thanks for your fine set of Simla Mosses, which are a welcome addition to my collection, and will afford me much entertainment in examining and arranging.

Was there ever such wonderful weather at this season?—such delicious spring days and such a steady west wind, yet without rain. Our rooks are building ; Daffodils and Ranunculus ficaria in

* From his office under Government of Inspector of Factories.

1859. blossom : Crocuses in full beauty, and Violets in profusion.   It is weather that tempts one out-of-doors, and consequently is unfavourable for study. What think you of the new Reform Bill ?   I like many things in it, especially the educational franchise ; but I doubt whether the ministry will be ably to carry it.   It seems likely as far as I can judge, to be vehemently opposed by the Radicals, and very coldly supported by the Conservatives. Then we shall have a dissolution of parliament, and very likely no one measure of real practical use will be carried in the whole session.

Mr. Prescott* does seem indeed to have been a beautiful character, and an irreparable private and public loss.   I cannot help very much lamenting that he should not have lived to tell the story of the Spanish Armada : a glorious theme, to which he would have done justice.

Ever your very affectionate Brother,
C. J. F. Bunbury.

JOURNAL.

April 1st.

News of the division on the Reform Bill ; majority of thirty-nine against the Government.

The debate has been a long and curious one ; curious, because so many of the speakers took an unexpected line ; so many contradicted themselves in their speeches, coming to a conclusion at variance with what might be inferred from their arguments.

* Mr. Prescott, the American historian.

Much excellent speaking. Lytton Bulwer's speech 1859. was splendid; those of the Solicitor-General and Sir John Pakington very vigorous and powerful; Sir James Graham's on the other side, extremely power- ful and statesmanlike; Mr. Bright's very able, and *suspiciously* moderate; Lord Palmerston's ex- ceedingly artful. These were what struck me most. I am very sorry the question has been agitated. There is now an end of really useful practical measures for some time to come.

LETTERS.

Mildenhall,
April 3rd. 1859.

My dear Susan,

Having now finished the first volume of your Colletta (I have read it very slowly by the way) I must write you a few lines to express my ad- miration both of the perseverance with which you worked through such an undertaking, and of your success in it. The translation appears to me ex- tremely well done, free, clear and vigorous; indeed I like your style better than that of the original, as being less forced and artificial. The style of Tacitus, which Colletta so much affected, is not much to my taste. You have been particularly successful in rendering the two most interesting and striking portions of this first volume, namely, the descriptions of the earthquakes in Calabria, and the sad chapter relating to the fall of the republic and

1859. the cruelties of the restored Bourbons. I have marked here and there a sentence in which I think the turn of expression might be improved.

Your supplementary chapter also I think excellent. I was very glad to hear Rufini's* opinion of your book; it is very gratifying, as he is so good a judge, being himself an admirable writer of English. Altogether, I hope your Colletta will have a great success, which it well deserves.

I thoroughly agree with you in believing that the Congress will be a grand humbug, out of which no real good can come, I still think—and so does my Father—that a great European war must come before very long, even if it be staved off for this year—which I do not yet feel sure that it will be. But who will ultimately be the dupe ? Sardinia ? or England ?

I am reading Ellis's "Madagascar," in which I find much to entertain me. The people of Madagascar seem to be rapidly acquiring the nuisances of civilization ; they already have custom houses and quarantine : I daresay they will soon have poor laws and petty sessions, perhaps even parliamentary elections ! Seriously, Ellis's account of them shows a very odd medley of barbarism with fragments of civilization.

I confess I tremble for the Superannuation Bill, whether the result of this Reform crisis be a dissolution of Parliament or a change of Ministry. I wish the Ministers had (as they originally intended) deferred their Reform Bill till after Easter, and carried

*Author of " Dr. Antonio," " Lorenzo Benoni " etc.

through various measures of homely practical utility 1859. before they plunged into that angry whirlpool. It is still possible that they may withdraw their bill, and keep their offices for another session; but I think it would be a course very damaging to their reputation. There were some very powerful speeches in the course of the long debate.

We have been in a considerable turmoil here for the last week, expecting a contest for West Suffolk.

My Father in spite of his fall, seemed very well when we were at Barton, and in very good spirits, and Lady Bunbury uncommonly well.

I look forward with great pleasure to seeing you here in the course of this month. With my best love to Mr. and Mrs. Horner, believe me,

Your very affectionate Brother,
C. J. F. Bunbury.

P.S.—Tuesday, April 5th. So Parliament *is* to be dissolved. I think the Ministers are quite right in doing so, but at the same time I am very sorry it should happen.

JOURNAL.

April 14th

Finished reading with Fanny the fourth book of the " Iliad." We have read together the third and fourth books since the 30th of March ; she reading Voss's German translation, which seems to be wonderfully literal, and I reading the original. In the evenings I am reading to her Froude's History of England. I had previously read to her in the

1859. evenings, since we have been settled at home, "Childe Harold" and "Beppo," and "As you like it," and a small part of Hume's 5th volume, which we abandoned for Froude.

_____

April 23rd.

The news of this morning shows war on the Continent to be now apparently inevitable : in fact it is all but actually begun. Austria has virtually declared war against Piedmont, wishing no doubt to gain the military advantage of the first blow.

_____

LETTERS.

Mildenhall,
Easter Sunday,
April 24th, 1859.

My Dear Katharine,

Very many thanks for your note and for the Asplenium marinum, which I duly received from Mrs. Horner. The specimens are superb; I wondered when I received them where you could have got them, as they had no label, and I supposed you were not likely to have gathered them either at Paris or in the Regents Park. They are quite like the Madeira specimens of marinum. It is curious that a plant which certainly inclines rather to warm than cold countries, and which is very tender in cultivation, should grow so luxuriantly in so northerly a latitude as the Isle of Eigg : but indeed I believe that the climate of the Hebrides is mild, though wet and stormy.

We have now, as you say, a good house-full, and 1859.
very pleasant it is. It is a pleasure to have
Leonora with us again, after so long an absence,
and a great satisfaction to see Mr. and Mrs. Horner
so well and cheerful: Susan is as usual, extremely
agreeable.

I wish the French armies may be in time to save
poor Turin. Whatever may be the ultimate results
of the war, it must certainly be attended with
such miseries and sufferings for poor Italy as one
cannot think of without horror. It is assuredly the
duty as well as the interest of England to remain
strictly neutral; but Lord Derby unfortunately
betrays such Austrian leanings, that the consideration of foreign affairs reconciles me considerably to the probability of *his* downfall.

On the 30th of March I found the Orthotrichum
striatum of Musc. Brit. (Orthotrichum *leiocarpum* of
Schimper) growing on an Ash tree in one of the
plantations: I had never seen it here before, and
could now find it only on one tree. This has been
a bad season for Mosses, for though, till lately,
remarkably mild, it has been as remarkably dry; so
that those Mosses which do not fruit *regularly*, such
as Hypnum Schreberi, Splendens, Squarrosum,
and Minum hornum, have been almost entirely
barren this season. Most of our greenhouse Ferns
are now in beauty, and unfolding their fronds
charmingly, such as Woodwardia radicans, Onychium
lucidum, Pellaea rotundifolia and Hastata; and
Ophioglossum vulgatum, which I brought in last
year from the swampy land towards Eriswell, and

1859. which has been kept in a pot in the greenhouse during the winter, is now coming up again and looks healthy. In our out-door fernery, Polypodium dryopteris has come up beautifully, but everything has been checked by this cold weather.

Ever your very affectionate Brother,
CHARLES J. F. BUNBURY.

JOURNAL.

May 2nd.

Attended the nomination of candidates for West Suffolk on the Angel Hill at Bury. Lord Jermyn escorted into Bury by a numerous cavalcade.

Lady Bunbury told me yesterday, that she had once met the famous Lady Hamilton in society in 1806 or '7 during the time of the "Talents" administration. Lady H. was then enormously fat, quite unwieldy; her eyes magnificent for size, colour and brilliancy, but with a very bad expression,—a hard, cruel, pitiless expression, giving the idea (Lady B. said) that the person to whom such eyes belonged would be capable of any deed of cruelty. Lady B. also told me that she knew on the authority of her friend Lord Northwick, that the active part which Lady Hamilton took in procuring the violation of the capitulation and the execution of Caracciolo, was determined by an actual bribe from the Queen of Naples, and that the money was actually paid to her on board Nelson's ship. Captain Foote, Nelson's flag-captain, remonstrated

earnestly against the breach of the capitulation, 1859.
till Nelson peremptorily silenced him ; and Captain
Foote was never employed again.

May 3rd.

The war has begun in earnest ; there is official
information that the Austrian army has entered
Piedmont and advanced as far as Novara : and
one account, which however does not seem to be
well authenticated, says that some sharp fighting
has already occurred, and that one Austrian regiment
has lost as many as thirteen officers. However
this may be, there can be no doubt that the war has
been fairly begun. Who can say when it may end ?
May God defend the right !

May 5th.

The Election for West Suffolk. Lord Jermyn
brought in by a great majority : nearly 600 ahead of
either of the other two candidates ; Major Parker
second, which I did not expect.

May 8th.

The newspapers announce the death of Humboldt,
after a short illness. It is a happy thing that
such a noble, active, and honoured life, prolonged
to so great an age, should thus have been terminated
without any lingering decay or loss of faculties and
with little suffering. It is a great pleasure to me to
remember my conversations with him in '55, when
I was at Berlin ; a great satisfaction to reflect that
I saw so much of him on that occasion, and

1859. especially to remember the kindness and cordiality that he showed me. The appearance and manner of the noble old man are vividly before me. Though he had lived so long, and done his work so well, and his death could not be altogether unlooked for: one cannot but feel sorrow that so bright a light of science should be extinguished. It will be long before the world sees another Humboldt. I am told that he himself said at the beginning of this year, that he knew he should not live to be ninety. He would have completed his ninetieth year if he had lived to next September.

Lord Bristol who was born the same year with Humboldt, died last February, less than three months before him.

LETTERS.

Mildenhall,
May 11th, 1859.

My dear Lyell,

The election and some parish botherations have rather delayed my reading of your paper on Etna; but I have now finished it and must congratulate you heartily on having added such a splendid wreath to your geological laurels. It is a masterpiece, and will I think give the death blow to Elie de Beaumont's hypotheses. In your first part, the proof of the formation of solid and stony lavas on steep slopes, in opposition to Elie de Beaumont's notions, is worked out with almost

mathematical clearness and precision. I do not 1859.
think I ever read anything in geology more con-
vincing. Your description of the eruption of 1852,
and of the changes produced by it in the appearance of the Val del Bove, is remarkably interesting.
In the second part, I was particularly struck by
the arguments relating to the double volcanic axis.
These are to me the most important and striking
parts of the paper. I hope it will become widely
known. If you have copies to spare, I think my
Father would like to have one, as he knew Etna
so well half a century ago, and I am sure he
would be interested by your description of the
eruption of 1852.

So that noble old man, Humboldt is gone at
last ; happy in such a speedy and easy end to
so long and so glorious a life ; preserving his fine
faculties to the last. " The general favourite and
the general friend."

I reflect with great pleasure on the conversations
I had with him at Berlin, four years ago, and the
genial kindness and attention that he showed me.

Your tour in Holland, short and rapid as it was,
must have been interesting and pleasant, and now I
suppose from your letters, you are hard at work
again—indefatigable man that you are.

I do hope you and dear Mary will be able to come
and see us in the course of the summer. We are
very sorry to part with Mr. and Mrs. Horner and
Susan, whose visit was extremely pleasant to us ;
but I wish Susan were stronger. The season
however seems to effect most people more or less.

1859.   I am working on steadily at the Indian Fossil
plants, a work of great detail; and am reading
Von Buch's "Canarischen Inseln," and Froude's
History of England. I am not yet convinced
that Henry the Eighth was a model king.

We had a pleasant two days visit from Mr.
Zincke; he is an agreeable and a valuable man,
of remarkable activity of mind,

Much love to dear Mary.

Ever affectionately yours,
CHARLES J. F. BUNBURY.

JOURNAL,

May 14th.

A letter from G. Pertz to Leonora (who is staying
with us) gives an interesting account of the death
and funeral of Humboldt. His death seems to have
been most peaceful and painless; a happy and
suitable close to such an illustrious life. The
honour paid to his remains by the general and
spontaneous feeling of the people of Berlin, is an
honour to the nation.

LETTERS.

From his Father.

Barton,
May 17th, 1859.

My dear Charles,

Your last and my last letter crossed
upon the road, so you have had no answer to the
questions you put to me about the war. I think

that the first view of the Austrians had been to 1859.
fall with superior numbers upon any bodies of
the Piedmontese who might have remained in
the plains, and thus to weaken the strength of
their opponents. But the Piedmontese had wisely
withdrawn to the hills and heavy rains cramped
the movements of the invaders. Then came into
play the second object of the Austrians. They
have taken advantage of the absence of the French
to subsist their army for a month at the expense
of the enemy and by ravaging the richest part
of Piedmont they have deprived the allies of the
resources of a country which would be the first base
of their operations. Now that the French have
arrived in force, I expect to see the Austrians retire
across the Ticino, and to wait for battle in positions
which they have long prepared. The general
situation of the Austrians is impaired by the
revolution in Tuscany and the increasing ferment in
Romagna and Central Italy. But the greatest of
the problems before us is this. Will Napoleon, who
has studied the theory of war all his life, prove to be
practically equal to the command of 150,000 men?
He has difficulties before him, and not the least is
that he must treat the Italians as friends, instead of
letting his army, as the first Napoleon did, plunder
and live at the expense of the country.

A fall of snow at two o'clock this morning, though
the thermometer was at 50 deg.

Much love to Fanny.

Most affectionately yours,

H. E. B.

JOURNAL.

May 23rd.

News of the first serious encounter between the French and Austrians; a sharp partial engagement at Montebello, the same place from which Lannes took his title. The Austrians retreated. It was, I presume, a reconnoissance in force; we have no satisfactory account of the Austrian loss, except that they left a colonel and 200 other prisoners in the hands of the French; but the fighting must have been severe, as the French lost a major-general, killed, and three colonels wounded, besides other officers. The Austrians are said to have been 15,000 strong, the force of the allies actually engaged is not stated, but a French infantry division and the Sardinian cavalry, appear to have borne the brunt of the action.

May 29th.

We have now some information though not yet very clear or full, as to the battle of Montebello, which was evidently very severe though indecisive. The Austrian official report admits a loss of 290 killed, 718 wounded, 283 missing: total 1291. French loss not yet officially stated. The total force of the Austrian corps engaged, is stated in their own official report at 25,000 men; that of the French and Piedmontese is estimated by their enemies at 40,000: while some of the French accounts make it only 5,000. These accounts are not irreconcilable: the French statement no doubt

refers to the number actually under fire, the Austrian 1859.
to the whole of Baraguay d'Hillier's corps which
came up in support. My Father writes to me :—
" My view of the severe action at Montebello is
" this. The Austrians thought that by falling
" suddenly, and with superior forces on Forey's
" division they might crush it. They succeeded
" partially, but they found to their disappointment,
" that Baraguay d'Hillier's corps d'Armée had
" arrived, and they then retreated, but were not
" pursued. The French appear to have had the
" worse of the fight : and their loss has probably
" been greater than that of their assailants. I
" expect to hear of a battle on a larger scale about
" Stradella."   It appears that the Austrians lost ten
officers killed and sixteen wounded,—a large proportion to their total loss.

There are also accounts, but not yet from
thoroughly authentic sources of successes gained by
Garibaldi over the Austrians in the country between
the Lago Maggiore and Lago di Como.

Dr. Pertz told us the other day that in the
campaign of 1814, the resolution of the allies to
march upon Paris, which brought the war to a
decisive crisis, was entirely owing to the Emperor
Alexander, instigated by the Prussian general
Gneisenau and Baron Stein ; that the Emperor of
Austria wished to make peace with France,
Metternich had the same wish, and Schwarzenberg
was governed by Metternich ; Lord Castlereagh and
his brother were likewise for peace, on the condition
that France should have for her frontiers the Rhine

1859. and the Alps; the King of Prussia was guided entirely by the Emperor Alexander ; and Alexander was influenced principally by Stein and Gneisenau.

Pertz has a very strong and settled feeling of aversion to the French ; not surprising as he remembers French soldiers quartered for many years in his father's house at Hanover ; but unfortunately, at the present crisis, this feeling makes him lean much to the side of Austria.

June 2nd.

News of fresh battles, and very sharp ones,—this time on the Sesia, therefore I suppose towards the centre of the Allies line. The telegraphic accounts are rather confusing, but as far as I can make out, the King of Sardinia, with his own corps d'Armée, crossed the Sesia a few days ago, attacked an entrenched position of the Austrians at Palestro (a place I do not find in the maps), and got possession of it, taking 2 guns and many prisoners. On the 31st of May, the Austrians, 25,000 strong, endeavoured to retake Palestro, and there seems to have been a severe action in which the Piedmontese were again victorious, taking eight guns and one thousand prisoners. These are all the particulars we have as yet, but as the advance of the King's corps seems to have been part of a general plan, we shall probably soon hear of still greater engagements.

Mr. Zincke, in a letter I had from him this morning, says of Humboldt very justly : " Humboldt's " life has more the aspect of the working out and

" realization of an idea than that of an ordinary life. 1859.
" It appears to have been very little distracted or
" disturbed by the circumstances and occurrences
" which affect other men's careers."

LETTERS.

Mildenhall,
June 8th, 1859.

My Dear Katharine,

We returned yesterday from Barton, where
we left my Father tolerably well.   Cecilia has not
been well these last days, but her children are in
blooming health, and very charming little creatures
indeed ; the two elder boys quite beautiful, to my
thinking, and the baby now thriving wonderfully.
Little Emmy very delicate and pretty, very odd and
clever, and amusing—a thorough Napier.   Little
Henry, an affectionate, warm-hearted little fellow,
quite devoted himself to Fanny, and took complete
possession of her.   I am very glad to hear so good
an account as you give me of your dear children.
We have glorious weather, and our garden is in
great beauty now; our roses coming out famously;
and I wish you could see our out-door Fernery, where
the Ferns are flourishing as if in a Devonshire lane,
or on the edge of a wood in Wales.   Barton too is
in its glory, the warm, damp weather has brought
out the foliage and flowers in almost overwhelming
luxuriance.   That inclosed and wooded country
looks really beautiful (flat as it is) at this season,
when the greens of the hedges and woods and

4 M

1859. meadows and young corn are so brilliant, and the
cottage gardens bright with flowers.

Lady Bunbury drove me yesterday to see the
wilderness of Rhododendrons (I do not know what
else to call it) at Rougham, about three miles from
Barton. It is exceedingly beautiful. The variety
of kinds and the size of the particular plants, not
equal to what one sees in the Regents Park
exhibitions; but one drives for half-an-hour
and more through the midst of them; and
they grow with all the picturesque beauty of
wild plants, *here* in masses and *there* scattered,
mingling with Fern and Broom, and fine old Scotch
Fir trees; the effect is charming.

We are as you may well suppose, deeply interested
in the Italian war. I heartily rejoice that the
Italian troops have played such a splendid part in
the fighting, as well under Victor Emmanuel as
under that fine fellow Garibaldi. I hope the news
we have had this morning is true as to the
evacuation of Milan by the Austrians. I shall be
delighted if they have given up that beautiful city
without damaging it.

But I cannot believe that the victory at Magenta
has been at all so decisive as the *Express* supposes, or
that the Austrians are at all likely yet a while to
make peace or to relax their clutch of Italy. We
shall not hear in a hurry of the surrender of Pavia
or Piacenza. My Father says, and I fully believe
him—that there will be a long campaign and many
a bloody battle before the French reach the Mincio.
I still trust that the end will be the liberation of

Italy, but I do not think the end is very near. If 1859. attained, however, it will be worth all the bloodshed and suffering.

I am reading Froude's History. He has an extraordinary power of narrative; equal to Macaulay himself in the art of engaging one's attention and interest in everything he relates; but his ways of thinking are rather extraordinary. If I ever read anything thoroughly *Jacobinical*, it is Froude's reasoning on the condemnation of Sir Thomas Moore.

My hearty love to your husband and children, and to all the party at number 17.

Ever your very affectionate Brother,
CHARLES J. F. BUNBURY.

JOURNAL.

June 8th.

Events now come in quick succession on the theatre of war; for the allied armies which had so long remained nearly stationary, collecting supplies and completing their arrangements, are now advancing in good earnest. The two severe actions at Palestro, in which the King of Sardinia's corps successfully engaged the Austrians, were on the 30th and 31st of May; and his movement was as I had supposed, part of a general advance of the allies. The Austrians retired from Novara and Mortara, and gradually, as it seems, from all the country west of the Ticino; and the main French army crossed that river; but on Saturday, the 4th, they were attacked by two strong Austrian Corps d'armée

1859. at Magenta, about two miles east of Buffalora (on the direct road from Novara to Milan), and there was a sanguinary battle. The Austrian accounts represent it as a drawn battle, the French claim a victory; at any rate the loss on both sides was probably heavy.

To-day it is announced that the Austrians have withdrawn from Milan, and that King Victor Emmanuel has been proclaimed there. I heartily hope that this may be true, and that the Austrians have really quitted that beautiful city without injuring it. They are probably falling back on their strong defensive line of Pavia and Piacenza, great fortresses which the French must besiege in form.

We have as yet no full account of the two battles of Palestro, but it is quite clear that the King of Sardinia and his Italians behaved with distinguished valour. Palestro, which is not in Arrowsmith's map is a village a few miles south-east of Vercelli, on the road to Mortara.

It would appear that the first design of the French had been to operate by the right bank of the Po, and to cross that river below the junction of the Ticino. At the time of the battle of Montebello, their principal corps were massed about Tortona and Voghera. But either finding the Austrian position too strong, or for some other reason, they countermarched, moved all their corps to the left of the Po, and have been operating by the line of Vercelli.

My Father thinks they have shown great

skill in effecting this change in their line of 1859.
operations.

It seems the Austrians have really quitted Milan,
at which I heartily rejoice. The accounts of the
battle of Magenta are most contradictory; all one
can make out is, that it was a very obstinate and
bloody fight, honourable to the courage of both
sides; that the French had the best of it, but not a
decisive victory.

I attended a great meeting of the Magistrates of
the County at Ipswich, convened by the Lord
Lieutenant (Lord Stradbroke) for the purpose of
discussing and making arrangements for the
formation of local Rifle corps. Met many acquaintances.

News of the overthrow of Lord Derby's Ministry:
a majority of thirteen against them in a very full
house, on a question of "want of confidence."
Parties being so nearly balanced, it will be very
difficult to carry on any Government.

The campaign has been much more rapid than
we had expected. The Austrians have been
retreating ever since the battle of Magenta, and
seem to be really falling back to the Mincio without
any further resistance; they have evacuated Pavia
and Piacenza, which were understood to be very

1859. strong fortresses, have withdrawn their garrisons from Bologna and Ancona, and even from Ferrara, and seem to be concentrating all their forces in the old Venetian and Mantuan territory. It is difficult to understand this, except indeed that the incapacity of Gyulai to cope with the French Generals has become manifest. My Father writes "that Gyulai " has made a miserable mess of his campaign. " Blunder after blunder! That the Austrian " soldiers have fought well there can be no doubt."

All the accounts agree that Magenta was a very obstinate and bloody battle : the village of Magenta itself is said to have been taken and retaken seven times, and probably the loss on both sides was heavier than is acknowledged. Now, as the allies have got possession so easily and speedily of that fine rich plain of Lombardy, they will be able to feed their army more easily than was at first anticipated, and I have little doubt of their ultimate success, if the war continues to be confined to Italy, although the fortresses of Mantua, Peschiera and Verona may stop them for some considerable time.

June 18th.

Finished reading "Adam Bede," a most interesting and indeed excellent novel, though somewhat dis- figured by the too frequent introduction of a barbarous provincial dialect.

The new Ministry is formed with the perennial Lord Palmerston for Prime Minister, Lord John

Russell for Foreign Secretary, and Gladstone 1859.
Chancellor of the Exchequer. It seems to be on
the whole a good set of names ; a few too democratic
to please me.

LETTERS.

Mildenhall,
June 19th, 1859.

My Dear Susan.

Let me congratulate you on the wonderful
progress which the allies are making, and the
liberation of so great a part of Italy from the
Austrians. I can well imagine what delight you
must be in. The campaign certainly has been
astonishingly rapid : far beyond what I dreamed of :
nor can I think that it would have been so
without great blunders on the part of the Austrian
generals, for their soldiers appear to have fought
well. But if it be true, as I see stated, that at
Magenta, whole regiments of Hungarians laid down
their arms almost without fighting, this disaffection
will shake the very basis of the Austrian power. I
expect that the allies will find serious obstacles in
the great fortresses on the Mincio and Adige :
and they may be brought to a standstill for a considerable time, but I do not believe that the
Austrians will ever regain their hold on Lombardy
proper. The Austrian tendency may be strong
among the fashionable people in London, but it is
evident from the dispatches now published, that
even Lord Malmsbury felt that the real public

1859. opinion of England was by no means in that direction. It is amusing to see how the *Times* is turning round, now that Austria appears to be losing the game.

The Emperor Napoleon promises well : nothing can be better than his manifesto : but a man who has once committed such an enormous act of treachery as the coup d'étât must pay the penalty in being for ever distrusted.

I wish you could all be here to see our roses, which are really glorious. We quite live among them. We are like Adam and Eve, alone in our garden, but I take it that Adam was not much troubled with parish business, nor Eve with house-keeping.

With very much love to all your party at Queen's Road West, I am your very affectionate Brother,

CHARLES J. F. BUNBURY.

————

Mildenhall,
June 23rd, 1859.

My Dear Katharine,

Very many thanks for your letter, and for sending me Hooker's to read, with its interesting information about Ferns.

I really admire your zeal and energy in finding time amidst all your family and household cares to work at Botany. The Struthiopteris I have never seen, and it will be an especial acquisition. I have no specimen even of the long known Struthiopteris germanica, and have never seen it except in the Botanic Garden at Edinburgh. My collection is particularly poor in European (non-British) Ferns.

I wish you could see our roses : they have been 1859. quite charming this month, and indeed are so still, though perhaps the very best of them is over. It would really have cost us a pang to leave them in their beauty. Our Ferns too are are very flourishing, and would I think please you : I wish you could see them.

If we are able to come to London, I shall be delighted to go to Kew with you, and to look through your Ferns.

Pray give Susan my best thanks for her very interesting political letter. We agree wonderfully well on politics at present ; the reason is, I take it, that this Italian question is not a question between aristocracy and democracy, but between despotism and free government. What especially delights me is the brilliant career and success of Garibaldi. I expected that we should have heard of a great battle about the position of Castiglione and Lonato ; but it seems the Austrians cannot make up their minds to try a decisive struggle there, and are retiring altogether behind the Mincio. What do you think of the chances of the present Ministry ? How long will they hold together ? How is it possible that Sydney Herbert and Gladstone can act satisfactorily with Milner Gibson and Cobden?

Pray give my love to your husband and your dear little people.

Ever your very affectionate Brother,
C. J. F. BUNBURY.

JOURNAL.

1859. News of a tremendous battle fought on the 24th, a little to the W. of the Mincio : the Austrians again defeated. There are as yet, of course, very few particulars : but it must have been a great and terrible battle, for the whole or nearly the whole of both armies seems to have been engaged, the Austrian line of battle is said to have extended five leagues in length, and the action lasted from 4 a.m. to 8 p.m. With such numbers engaged, and for so long a time the slaughter must have been frightful ; indeed the Austrian account admits that *their* losses were extraordinarily heavy. It appears that the Austrian army, after having retired to the E. or left bank of the Mincio, crossed again to the W. of that river on the 23rd, headed by their Emperor in person, attacked the Allies in their positions between the Mincio and the Chiese, and being defeated, recrossed the river in retreat. Solferino, where the especial brunt of the battle appears to have been, is very near Castiglione, where the first Napoleon gained a victory in 1796.

News that the whole French Army has crossed the Mincio, and that the Piedmontese have begun the siege of Peschiera. There are now also some fuller accounts of the battle of Solferino (on the 24th of June) ;— a terrible and bloody battle it was, though not to be ranked with those of Borodino or Leipsic. The French state their loss at 720 1859. officers and 12,000 men, killed or wounded; the Austrians *theirs* at about 10,000* killed and wounded. A fearful carnage.

———

To London by the early train—Fanny accompanying me as far as Newmarket. A lovely morning. Arrived at 17, Queen's Road West—most kindly and cordially welcomed by my dear friends there. In the afternoon, went with Susan and Joanna to the Zoological Gardens, where a great assemblage of gay company, enjoying the shade and the beautiful weather, were listening to the band. Met Charles and Mary Lyell, Katharine and her beautiful children, the Pulskys, the Edward Romillys, Henry Mallets, Lord Enniskillen, also Agassiz, who is now on a visit to England. His appearance is not remarkable. I had not the opportunity of conversing with him. Afterwards (from 6 to half-past 7), a small party at Mr. Horner's, where, in addition to those mentioned, were Marianne Galton, Mrs. William Nicholson, the Louis Mallets, and others. Mr. Pulsky in high spirits; not uneasy on account of the armistice which has been so suddenly concluded between the belligerents in North Italy; does not think it will be disadvantageous to the Italo-French allies, but that on the contrary it will allow time for the warlike ardour of the Germans to cool.

* This, it now appears, relates to only *one* of the two Austrian armies engaged.

1859.   Talked politics with Louis Mallet;—he does not
think the present ministry will last long, or has any
elements of duration in it; besides the political and
personal differences between Lord Palmerston and
Lord John Russell (suppressed for the present, but
almost certain to break out sooner or later), there
are vast differences—affecting the very first principles of government—between Lord Palmerston
and the Democrats. Louis Mallet thinks, and I
quite agree with him, that it would be a much better
and more natural coalition, if Lord Palmerston
and some of the more moderate of the present
ministry were to combine with some of the more
moderate of the late Government, such as Sir John
Pakington and other Liberal-conservatives; and
that this is not unlikely to come to pass in a year or
two.

July 10th.

Visited dear Minnie, saw her and Lady Napier
and charming Sarah; stayed to luncheon with
them, and had a comfortable chat. Then to Katharine; spent an hour-and-a-half with her very
pleasantly, looking over her fine collection of dried
Ferns from Assam, and talking over subjects connected therewith. Aftewards visited the Zoological
Gardens with Joanna and Leonard. Saw the two
hippopotami, both in the water at once; the
capybara (or cabiai) a recent acquisition from South
America, very quiet, tame, and stupid looking,
reminding me of Humboldt's account of it in his
Travels;—two young leopard cubs, very beautiful

and active little creatures. In one of the tanks in 1859.
the fish-house are some living *hippocampi*, most
singular little fishes with slender prehensile whip-
like tails, which they coil round any object in their
way, and with the head and neck looking like
grotesque caricatures of those of a horse in miniature.
In another tank some small fresh-water tortoises,
rather pretty, and very active in the water.

July 11th.

A brilliant and intensely hot day. To Kew with
Katharine,—spent some hours there very agreeably.
The gardens in exceeding beauty : the velvet green
turf (brilliantly green in spite of the heat), the shade
of the noble trees, the lake and fountain most
refreshing in such weather : the flower beds
splendid. We went carefully through the principal
Fern house : a magnificent collection, the plants in
the finest order and condition, their beauty and
variety wonderful.

In one house we saw the curious Lattice-leaf
plant, the Ouvirandra fenestralis, from Madagascar ;
the leaves of this grow entirely submerged, which I
had not understood from Ellis's account.

We missed Joseph Hooker in the Gardens, and
had only some ten minutes talk with him at last in
his own house. Talking of trees, he said the finest
trees he had seen in India, were Figs and
Terminalias : excepting these, he did not think that
really fine trees — single trees, were nearly as
common in India as in England.

July 12th.

Heat excessive. Visited the British Institution, where there are several excellent pictures by Gainsborough, both portraits and landscapes; in particular a view on the Suffolk coast, with a stormy sea, very remarkable for its truth to nature, and for the striking effect produced where the elements of the picture are simple and commonplace. Noticed also in the same exhibition, two striking landscapes by Salvator Rosa, with figures; the subjects from the story of Polycrates. In the one where Polycrates is crucified, there is a savage ghastly character given to the landscape, wonderfully suitable to the subject. Saw also a large picture now exhibiting in Bond Street—" The Heart of the Andes," by an American artist of the name of Church. A river and deep rocky ravine in the foreground, with rich tropical vegetation,—Palms, Tree Ferns, woody climbers and rich masses of blossom, with parrots and other brilliant coloured birds : all the details very well painted, with great truth to nature, and showing the character, not of the greatest forests of the torrid region, but of the first stage of elevation, the region of Ferns, 2 or 3,000 feet above the sea level. In the middle distance, a broad valley and wooded mountains rising beyond it with a good effect of distance ; and lastly, in the background, a vast snowy peak.

July 13th.

Had luncheon with the Charles Lyells, and a

pleasant chat. Then to the Athenæum, and had a 1859.
long talk with Edward.

Everybody is talking of the astounding *peace*, the
news of which came this morning. Never was
anything so unexpected : never such a falling off as
in this peace compared with the Emperor Napoleon's
proclamation from Milan, only a month ago. It is
a stunning blow to the hopes of the friends of
Italy. Peace settled in a few hours private conversation between the two Emperors, without
consulting a minister : the King of Sardinia absolutely ignored by his Imperial ally, and not even
invited to be present at the conference ; the *Pope* to
be the " honorary President " of an Italian
Confederation ; Tuscany, Modena, and Parma to
be restored to their expelled princes ;—can anything
be more despotic in manner, or more unsatisfactory
in substance ? Lombardy it is true, as far as the
Mincio, is liberated from the Austrians, but the way
in which even this is done appears to me offen-
sively despotic and dangerous as a precedent ;—the
Emperor of Austria cedes *his rights* over Lombardy
to the Emperor of the French, who transfers them
to his ally of Sardinia, without the slightest allusion
to any claim on the part of the people of the country
to have a voice in the matter.

July 14th.

Weather slightly cooler. Went to Clapham ;
saw my Aunt and Cousins, but not Sir William,
who is very ill. At the Athenæum, saw in the
newspapers the resignation of Cavour, — which

1859. clearly shows *his* opinion of the peace, and the Emperor's proclamation to his army. The expression in this, indicating what he wished to be considered as his motive for concluding peace, is remarkable :—" the struggle threatened to assume " proportions not in relation to the interests which " France had in it." — This seems to point to Germany.

Dined with the Charles Lyells. The party : Mr. Motley, the historian of the Dutch Republic, his Wife and Daughter, Mr. Hillard (another American author of a pleasing book on Italy), Erasmus Darwin, Mr. and Mrs. Horner and myself. Mr. Hillard talked a good deal and well. A small party in the evening : Professor Cappellini, a young naturalist from Pisa, with whom I had some talk ;— he seems zealous and well-informed on the state of science, but being one of the Liberal party, is much dejected on account of the recent news ; Sir James Lacaita, Sir James Clarke, Susan and Joanna, Edward and a few others. Mr. Hillard said that he found in England the highest and the lowest specimens of civilized humanity ; that the most cultivated and enlightened Englishmen are superior to anything to be met with in America or elsewhere, but the lowest class in London are more rude, ignorant, and brutal than any class to be found elsewhere in the civilized world. He remarked the striking rapidity with which the lowest and most barbarous Irish immigrants into America are assimilated with the general population, — how quickly they are improved by their altered circumstances and by the example of the people who 1859. surround them.

Charles Lyell's attention is much engaged at present by the curious observations lately made as to the occurrence of tools or weapons of flint, shaped by man in the gravel of the Drift period, and in caves, in company with (and in some instances even *below*) the remains of extinct animals.

July 15th.

Again extremely hot.

Visited the Water-colour Exhibition (the old Society), where there are many good and agreeable works. There is apt indeed to be too much same-ness in the works of these artists, but there are many very charming landscapes and sea-pieces, and some excellent figure pieces. Met Donaldson at the Athenæum, and with him the two Schlagintweits, those enterprising young men, Humboldt's protégés, who have penetrated further into Central Asia than any one before them. The third brother whom I had met some years ago at the Geological Club, has been murdered in the interior of Asia.

Met Mr. and Mrs. Motley again at dinner at Mr. Horner's, also M. Alexandre Prevost from Geneva, his Son, and Erasmus Darwin. Mr. Motley *came out* more than the former evening, and talked well on the school system in America. He has not the downright simplicity of manners that I have observed in some other eminent Americans, but has something of the European fine gentleman.

1859.   July 16th.

Returned home ; the heat excessive.

————

July 22nd.

My Father writes to me on the 18th, that he fully
concurs with me in what I say about the peace and
about Louis Napoleon :—" With regard to the latter
" I go perhaps farther than you may do, for I must
" turn to the histories of Asiatic tyrants to find an
" equal instance of perfidy and treachery. One, and
" only one, good effect this may have. All nations
" must now see that his professions are not to be
" trusted for a moment, and the more friendship
" he pretends the more necessary it is to stand
" vigilantly on one's guard. This dangerous man
" has now attached the French army to his person,
" and he has won for himself a high military reputa-
" tion. If he had fulfilled his promises and pro-
" fessions with regard to Italy, he would have gained
" and would have deserved the purest and noblest
" fame ; but his ambition springs from a mean,
" crafty, and unscrupulous selfishness. I wish I
" could agree with Lyell about Lombardy, but I
" think it a sham. It will not surprise me at all
" to see this bastard Napoleon pick a quarrel with
" his late ally and dupe, over-run Piedmont and
" Lombardy and annex both to France. It would
" answer his purpose as to an Italian Confedera-
" tion.

————

July 24th.

There is something peculiar in the tone of the

Emperor Napoleon's addresses to the Ministers and 1859.
to public bodies since his return to Paris ; a tone
not only of dissatisfaction, but almost of apology
rather than of triumph or exultation. He appears to
be much out of humour. He speaks of the difficulties in the way of his advance into Venetia, that the
military resistance of the Austrian fortresses could
have been overcome only by invading neutral territory in order to turn them, and thus entering into
a war with other powers and allying himself frankly
with the cause of revolution. He speaks of the
injustice of Europe towards him, and that he could
not have accomplished his original plan with regard
to Italy without embarking in a general war, *contre
le gré de l'Europe*. I still believe that one principal
reason for his making peace so abruptly was the
difficult and false position in which he found
himself placed with regard to the Pope ; that he
could not abandon the Pope for fear of the priests
and Jesuits in France, and could not support him
against his own subjects without flagrantly belying
all the pretexts of the war. This contradiction was
evident enough from the first, and the massacre at
Perugia must have forced it on Nap's attention.

LETTERS.

Mildenhall,
July 27th, 1859.

My Dear Mr. Horner,

    I hope you and Mrs. Horner will enjoy
your visit to Edinburgh, and will return to the
South in good health.

1859.   Do you ever remember a time when the political
world was more uneasy, more perturbed, more
uncertain and anxious?—when there was a more
universal feeling of insecurity, of dissatisfaction, of
vague alarm, of apprehension of some impending
evil without knowing exactly what? People seem
more uneasy in this state of pretended peace than
during the actual war. Everybody seems as if
asking, *where* the next storm will burst! Mr.
Bright appears to be the one solitary exception.

I am ever,

Your affectionate Son-in-law,

CHARLES J. F. BUNBURY.

JOURNAL.

August 8th.

Read General Shaw Kennedy's pamphlet on the
Defences of Great Britain, lent me by my Father.
It gives a striking picture of the insecure and
defenceless state of this country, supposing an
enemy once landed, and a very alarming view of the
ease with which a large French force might be
landed on our coast. My Father (who is here with
us at present) thinks the plan of defence proposed
in this pamphlet unsound, and indeed impracticable
in as far as relates to the proposal of surrounding
London with a circle of detached forts, at a mile
apart; he thinks that independently of the expense,
the political jealousy and suspicious vigilance of the
English would never tolerate such a measure, and

that, even in a military point of view, its advantage 1859.
is doubtful. My Father is, however, of opinion
that our principal arsenals absolutely require to be
protected by thorough systems of fortification : and
in particular, that Woolwich is by no means
properly secured, and that its locality is not a safe
one for our greatest military arsenal. He would
have our artillery and military stores distributed
among *two* other principal arsenals,—one at or near
Ely, the other near Bristol.

August 11th.

Mr. Bowyer, the Union school inspector, left us,
after a short visit, having come on the 9th. He is a
very clever and remarkably agreeable man, of
great information and great variety of conversation,
and very pleasant manners ; his political opinions
very different from those of his wrong-headed
brother. He thinks that the actual termination of
the war is on the whole more promising to the
Italian cause, than the continuance of the war and
the complete success of the French arms would
have been, for the Sardinian army would have been
exhausted in the long and sanguinary struggle which
must have preceded such a success, and the
Italians would have been brought absolutely under
French influence by their gratitude and enthusiasm
as well as by the necessities of their position.

August 12th.

We went up to London by the early train. Mr.
and Mrs. Horner were absent in the north, but

1859. we had a very pleasant chat with Susan and Joanna. I went to the French Gallery in Pall Mall, and saw Madame Henriette Brown's picture, of " Two Sisters of Charity Nursing a Sick Child," which Susan had mentioned to me. It is a very remarkable picture : the subject is to me not pleasing, but the treatment combines the exquisite finish and truth of detail of the old Dutch painters with much expression and feeling. In finish and truth of detail, this lady is at least equal to the best of our Pre-Raphælites, without their hardness and studied affectation of ugliness. The expression of the sick child—not the expression of the face only, but the whole disposition of the limbs is natural and pathetic : and the force of relief with which the child's limbs are brought out against the nurse's dress, is striking. Susan remarked to me, too, and very truly, a delicate shade of distinction in the expression of the Sister's face who is nursing the child, that kind and gentle and tender as it is, it is *not* the expression of a *mother*,—for it is not anxious.

August 13th.

Went down in the afternoon to Bognor, to see Emily (William) Napier and her children.

August 15th.

From Bognor to Brighton.

During this week at Brighton, the weather being very beautiful, I was almost constantly in the open air.

We read a novel, or rather collection of tales— " Scenes of Clerical Life," by the authoress of Adam Bede. The first of the three stories "Amos Barton " is very pretty. The second " Mr. Gilfil's Love Story "—really beautiful ; to me indeed more interesting even than Adam Bede. The third, " Janet's Repentance " I do not like at all.

On the 22nd we made an interesting excursion, in company with Sir John Boileau (who is a skilful antiquarian and an agreeable man), to Pevensey and Hurstmonceux.

From the comparatively high ground within the walls, we had an extensive view over the wide dead flat, marshy, alluvial level (no doubt formerly sea), which nearly surrounds the castle : bounded on one side by the high chalk downs running out to Beachy Head, on another by the uplands behind Hastings, and on a third by the line of the sea shore, with its series of Martello towers. The site of Pevensey must originally have been a peninsula, if not an actual island.

Thence we had a pleasant drive of five or six miles to Hurstmonceux. The castle here is a very noble and beautiful castellated house (or rather the shell of a house, for the interior is all destroyed) entirely of brick, of the time of Henry VI. The gate-house is especially fine : so is the E. side of the house as seen from the meadow, which was

1859. originally a pond, and afterwards (having been drained in Elizabeth's time) "The Pleasaunce." The Ivy is remarkably fine here as well as at Pevensey. There is a row of most noble old Spanish Chestnuts along-side the moat on the West side of the Castle. We had not time to examine details, nor to visit Hurstmonceux church, where Julius Hare is buried.

August 23rd.

From Brighton to London to Mr. Horner's.

August 26th.

We returned home.

LETTERS.

Mildenhall,
August 27th, 1859.

My Dear Katharine,

I received your pleasant and interesting letter a fortnight ago, when we passed through town on our way to Brighton, and I thank you very much for it. We spent only a week at Brighton, but it was a very pleasant holiday, and has done me a world of good. It is certainly a wonderful air. We had glorious weather, and I was gloriously idle : I bathed in the sea, lounged on the beach and the pier, drove out with Fanny and Edward, read little or nothing, wrote as little, and in short did nothing but imbibe the fresh air : and I am mightily the better for it. Doctor Brighton is for me one of the

best of doctors. I trust that Fanny also is really 1859.
the better for it: certainly she appeared so before
we left it. Edward spent part of every day with us:
he is not yet well, though gradually gaining ground;
he and Fanny talked away at a fine rate. The
only other person there whom we knew was Sir
John Boileau, with whom I was very glad to improve
my acquaintance: he was very agreeable. We
made a most interesting day's excursion with him to
Pevensey and Hurstmonceux: the Roman walls at
the former place (with the ruins of a Norman castle
within their enclosure) and the magnificent castellated house of the 15th century at the other, are
exceedingly well-worth seeing.

We had the pleasure of seeing Mr. and Mrs.
Horner, as well as Susan and Joanna, looking very
well before they set out for the Isle of Wight.

Ever your very affectionate Brother,
CHARLES J. F. BUNBURY.

JOURNAL.

September 10th.

During this fortnight, my primary study has been
the examination and description of the great
collection of fossil plants from Nagpore in Central
India, entrusted to me for that purpose by the
Geological Society. I have been working at them
little by little, ever since February, and have yet
much to do. I have been also going on with (what
I have had long in hand) the examination and

1859. cataloguing of all the plants in my herbarium from La Plata and Uruguay, with a view to a second paper on the subject for the Linnean Society.

I am now reading the correspondence of Charles Fox, edited by Lord John Russell : and I am reading Walter Scott's " Fair Maid of Perth " aloud to Fanny.

September 11th.

Read to Fanny Robertson's noble sermon on " The Message of the Church to Men of Wealth," that is, in fact, on the Christian view of the Rights of Property.

September 12th.

Very disagreeable news from China. The treaty has been broken by the Chinese themselves with the utmost treachery and unprovoked violence, and as we were unprepared for such a casualty, we have met with a serious disaster. It appears that the combined English and French squadron, having on board the ambassadors of the two nations on arriving at the mouth of the river Pei-ho, by which they were to ascend to Pekin, found it barred by stakes and defended by batteries strongly manned, which opened fire on our ships, and after a severe action, the squadron was obliged to retire, having lost 22 officers and 440 men, killed and wounded. Names not yet known. Thus we are plunged into another Chinese war, — one of which the justice cannot be disputed ; but as the enemy have shewn that no treaties will bind them, it is difficult to see what end the war can have. The only satisfactory point is, that the French are involved in it along with ourselves.

1859.

LETTERS.

Mildenhall,
September 19th. 1859.

My Dear Katharine,

Many thanks for your letter of the 13th. It is a great pleasure to hear such cheerful and comfortable accounts of all of you. My dear Father is, I am sorry to say, in a very serious state of health ; there is no immediate danger, and there is hope of recovery : but internal gout, continuing obstinately for near a fortnight, and not yielding as yet to any remedies,—and this in a man of 81, is enough to make one anxious. He suffers not much positive pain, but much difficulty of breathing, and appears very feeble, though not much altered in face. I returned from Barton yesterday. Edward and Henry are there.

If all goes well at Barton, I look forward with great delight to seeing Charles and Mary and Susan here next month.

That verdigris green colour in wood, such as you sent me a piece of, is produced by the spawn, (or, in technical language, the *mycelium*) of a small fungus called Peziza aeruginosa. I have seen it only in decayed wood, like your specimen, but Berkeley says that wood coloured by it is sometimes sound enough to be used in Tunbridge ware. I

1859. well know the pretty purple Clematis,—Clematis viticella, it is a native of Italy : we have a plant of it here, but being in a bad situation it does not flower.

Do you know the Pampas Grass, Gynerium argentum ? It is just now coming into flower at Barton: a fine tall stately silky panicle, like a reed, but silver white : the panicle and flower stem together between six and seven feet high. The beautiful Belladonna Lily is now in full bloom at Barton in an open border, looking exceedingly healthy, in a temperature very unlike that of its native Cape of Good Hope.

I saw in the *Express* a notice of Charles Lyell's discourse at Aberdeen, on the flint hatchets, &c., but I expect to see a better report of it in the Athenæum.

With much love to your husband and children,

I am ever your very affectionate Brother,

C. J. F. B.

JOURNAL.

September 23rd.

My Father's alarming state of health obliging me to go repeatedly to Barton has interfered with my journalizing as well as my other pursuits. He is now better, and I hope recovering. We have had full particulars of the disaster in China, and a very lamentable business it is in every point of view ; one of the severest checks we have ever met with in the East, with a sad slaughter of our brave fellows and such a blow to our *prestige* as is likely to do serious mischief in China, if not in India also. It is evident that our expedition was opposed by a much more skilful and formidable enemy than any we had before encountered in China ; that our force was insufficient to force a passage against serious resistance ; and that our people showed great bravery with no other effect than a useless sacrifice of their lives. It is announced that a joint expedition will be sent out by France and England to enforce satisfaction ; it is to be hoped that an ample force will be employed, for it is evident that the resistance and the difficulties will be serious. The affairs of Italy still continue in doubt and suspense, the policy and purpose of the French Emperor still dark and inscrutable. Whether he means to follow the just and noble course of insisting that the Italians shall be left free to settle their own affairs; whether he means to play into the hands of Austria, or whether (as is to be feared), his secret design is to cajole or bully "The Duchies" (Tuscany, Parma, and Modena), into choosing his cousin Prince Napoleon for their king;—is altogether a mystery. Nothing can be more admirable than the conduct of the Italians has been hitherto.

September 24th.

Yesterday I finished the first volume of the Correspondence of Charles Fox, and read to Fanny Prince Albert's Address to the British Association at Aberdeen.

1859.   I have very lately received the 14th number of
the Linnean Society's Journal, and read the President's (Bell's) Address, and the obituary.   Between
May 1858 and May 1859, the Society lost Robert
Brown, Humboldt, Bonpland, and Dawson Turner.

We finished the "Fair Maid of Perth," a few
days ago, and I have begun to read Tennyson's
"Idylls of the King," to Fanny in the evenings.
Read in the newspapers the account of Captain
McClintock's discoveries relating to the fate of
Franklin and his crew :—interesting and sad.

Poor Sir John Franklin died on the 11th June,
1847, and it was not till a year afterwards that the
then survivors of the expedition abandoned their
ships in the ice ; nine officers and fifteen men
had previously died of sickness ; and it is probable
that the remainder died of exhaustion, for it appears
that they did not want food.

October 1st.

In the early part of this week we were again much
alarmed about my Father, and I was again called to
Barton : but he has rallied, and there is, I trust,
great probability of his ultimate recovery.

Since I last returned from Barton, I have been
working at Lepidodendrons.   The Geological
Survey have sent me down all the fossil plants of
that genus in their collection to be named ; and I
have been busy examining, comparing and labelling
them.   The determination of them is very difficult,
owing to the variations of the plants themselves,

the further variations produced by fossilization, the 1859.
uncertainty as to the real value of characters, the
imperfect state of many specimens, and the incomplete and insufficient information to be found in
books. I have been working with all the care and
diligence I can use, but shall doubtless have committed many mistakes, for here, even more than in
palæobotany generally, I feel that we are at best
only groping our way.

In the second volume of the Correspondence of
Charles Fox, I have read as far as the formation
of the Coalition Ministry, on which subject Lord
John Russell's observations appear to me very
judicious. I have read partly aloud to Fanny, and
partly to myself, Tennyson's new poem, "The Idylls
of the King," which has great merit and is very free
from the obscurity that may be complained of in
many of his writings. The character of King
Arthur is very noble, and especially his last address
to Guinevere is pathetic and beautiful.

October 2nd.

Read two of Hume's Essays: the one on Poly-
gamy and Divorce, and that on National Character.
I am much more struck with the merits of his style
in these essays than in his history; there is a
charming ease, simplicity and clearness in it. In
the essay on National Character there are many
very judicious observations; he contends very forcibly against the hypothesis carried to such extravagant lengths by Montesquieu, that climate is
the chief agent in producing the varieties of national
character.

Mildenhall,
                              Sunday, October 2nd, 1859.

My Dear Katharine,

I thank you very much for your very kind
letter of the 27th. I always rely with thorough
confidence on your kind affectionate sympathy, both
in my joys and sorrows, and I can assure you it
is very dear and very precious to me. Indeed
I have few greater comforts in life than the constant
kindness and affection and sympathy which I have
always experienced from you and from all Fanny's
family.

The accounts I received of my Father last Monday made me very uneasy, and I went over to
Barton to see him. I found him thoroughly convinced that he was dying; perfectly calm, resigned,
I might say contented, and in a most happy state of
mind, and I had a conversation with him that
evening, which it will always be a pleasure and a
comfort to me to remember. Still, from his appearance, I could hardly believe that he was so
near death as he thought, and since then he has
rallied, and seems to be going on steadily well; so
that I hope and believe he will recover; though
of course his advanced age is a serious consideration.

The variable and treacherous weather has given
me a cold, but it is not very bad. The letters
we receive from all of you are among our greatest
entertainments.

The woods and shrubberies at Barton are now
very beautiful with the various bright colours of the

leaves ; they have changed earlier than here, and 1859.
the autumnal tints are remarkably fine this year,
especially on the American trees and creepers.

I look forward with great delight to seeing Susan
and Charles and Mary ; I am afraid we shall not
tempt you down, but must wait till we go to
London.

Ever your very affectionate Brother,
CHARLES J. F. BUNBURY.

JOURNAL.

October 6th.

I can no longer shut my eyes to the probability
that the end of my dear Father's life is near. He
has for some time past been firmly convinced,
himself, that he is dying, and though happily he
suffers little, he seems to have no clinging to life :
he is perfectly resigned, even contented to die, and
is in a most calm and happy frame of mind, feeling
that he has done his work, and done it well, and
that he has (as he expresses it) settled all his affairs,
religious and worldly.

God grant that my latter days may be like his.

LETTERS.

Mildenhall,
October 31st, 1859.

My Dear Katharine,

It is a long time since I have written to
you, but you are well aware what a time of sad

1859. anxiety and suspense we have had with my dear
Father's illness. He is yet far indeed from anything like decided recovery, but he has remained
now so long in the same state, without evident
improvement indeed, but equally without any visible
change for the worse, without positive suffering and
without any very positively bad or alarming
symptoms,—that our immediate apprehensions cannot but be removed, and we are returning to our
ordinary occupations and pursuits, though we can
form no definite plans for the future. I fear there is
little prospect of his recovering *completely*.—I mean
so as to be as well as before,—but, as far as we can
see, he may continue a long time in his present
state, or with some little improvement. As he
prefers being alone, and I can be of no present use
to him at Barton, we have returned home, and I
have returned with great zest to my museum and
my regular studies. You will have heard about us
from Charles and Mary. It was a very great
pleasure to me to see them, and to have some good
talk with them, though my enjoyment of their
company was sadly broken and interrupted. How
eager Charles Lyell is about the flint hatchets, and
about Darwin's forthcoming book on Species. This
book is indeed sure to be very curious and
important, and is likely to cause no little combustion
in the scientific world, for I have no doubt that
plenty of pens will be drawn on both sides of the
question. After all, however mortifying it may be
to think that our remote ancestors were jelly fishes,
it will not make much difference practically to

naturalists who deal with recent plants and animals; 1859.
for species must be distinguished and named,
whether we suppose them to have been distinct from
the beginning, or to have been produced by causes
of variation acting through enormous periods of
time.

I am now hard at work naming a collection of
Lepidodendrons entrusted to me for determination
by the Geological Survey; and awfully hard they
are to make out.

Dear Susan's company is a very great pleasure to
us: she is always so agreeable and so kind in
staying on with us, though we were obliged to leave
her so inhospitably. I have not time to write
more at present, so must say farewell, with much
love to all your party.

Ever your very affectionate Brother,
CHARLES J. F. BUNBURY.

JOURNAL.

November 26th.

To London; most kindly received by Harry and
Katharine Lyell. A cheerful family party in the
evening, to celebrate little Arthur's birthday.

November 27th.

Sunday.—Read Channing's admirable sermon on
the Character of Christ. Read also a chapter of
Darwin. Went to the Zoological Gardens with
Fanny, Joanna and others; saw the so-called wild

1859. Horse from Thibet—a new acquisition : it looks very like a mule, and like one that has more of the ass than of the horse in its composition.

November 28th.

Dined with Charles and Mary—no one else present ; a pleasant, chatty evening. Charles Lyell very full of Darwin's book and quite a convert to his theory, which will make great changes necessary in the next edition of the Principles.

November 29th.

Went with Charles Lyell to the Botanic Gardens in the Regents Park. A curious new plant in flower in the great conservatory,—has been in cultivation there for eleven years, and now flowering for the first time ;—not named ; the gardeners say that Sir William Hooker declared it to be unknown to him. It is evidently allied to Agave :—a long stout flower-stalk, rising from the midst of a huge tuft of radical leaves, bears a dense, pendulous spike five feet long, of innumerable close-set, pale green, bell-shaped flowers with long stamens. Not much else in flower in the house, except the Chrysanthemums, which are very fine, and two or three Australian Acacias. In the hothouse a very beautiful white-flowered Pancratium-like plant from Brazil, Eucharis Amazonica, with broad leaves like Hemerocallis cærulea ; also Ixora coccinea and Poinsettia.

In the great walk we met Babbage, who told us how he had just before been defeated at the police-office

in a contest with his old enemies the street mu- 1859.
sicians.

November 30th.

Went at 2 p.m. to meet Mr. Horner at the
Geological Society, where he showed me the excellent arrangement he has made of the rock
specimens and the catalogue he has drawn up of
them—a model of neatness and of clear and accurate information.

At the Athenæum, read a portion of Sir Emerson
Tennant's "Ceylon," which seems to be a valuable
and excellent book; the part I read, which is that
relating to mineralogy and botany is very well done
and full of good and interesting information.

Dined at the Geological Club :—a scanty attendance, owing to the anniversary meeting of the Royal
Society on the same evening; but I had some
pleasant talk with John Moore.

Evening meeting of the Geological Society. The
first paper, a notice by Mr. Atkinson, the Siberian
traveller, on some *bronze* ornaments found in the
auriferous drift of Eastern Siberia ;—a pretty good
discussion on this curious subject—much more
curious and extraordinary if the fact be well ascertained, than the flint knives or hatchets in the
European drift, as articles made of bronze imply a
much higher stage of civilization than those of flint.
In reality it is most probable that these Siberian
bronze articles are of much later date than the
auriferous drift, whatever might be the accident

1859. by which they were brought into the midst of it. I was glad to have the opportunity of seeing Atkinson, whose travels I had read with interest.

The second paper was a good account, by a Roman Catholic priest, of the volcanic cones and craters near Auckland in New Zealand ; a district in which these volcanic phenomena seem to come uncommonly thick.

Lastly, two papers on certain geological characteristics of a part of South Australia.

LETTERS.

Friday,
December 2nd.

My Dear Emily,

I am exceedingly sorry to learn that my dear Father has again been so very unwell, and that the comfortable progress which he appeared to be making has been interrupted, and that he has again had bad nights and suffering from oppressed breathing. It is sadly disheartening, and I feel very much for you, to whom it must be so especially painful to see him falling back into his former state of depression and discomfort. Still I hope that this may be only temporary and that he will rally again. The weather is terribly against him. If the accounts of my Father are pretty good, we mean to run down to Brighton this day week, probably for a fortnight, and afterwards to stay at Mr. Horner's till after New Year's day.

I am to meet Sir George Grey (of the Cape) at 1859.
dinner to-day ; I am extremely curious to see him.
I will afterwards write to you what I think of him.
I saw Mr. Atkinson the Siberian traveller the other
day at the Geological Society.

Pray give my best love to my dear Father, and
believe me,

Your affectionate Stepson,
CHARLES J. F. BUNBURY.

JOURNAL.

December 2nd.

Dinner party at Mr. Horner's : Sir George and
Lady Grey, the William Nicholsons and Graham
Moore Esmeade, besides the family. Sir George
Grey (the Governor of the Cape, and formerly of New
Zealand), I was very desirous to meet, having a high
admiration for his character and administrative
talents, besides having heard much of his agreeable
qualities. He comes up to my expectations. His conversation is not only agreeable, but gives the impress-
sion of a clear and powerful intellect, and of a strong
character. He spoke highly of the usefulness of the
missionaries in South Africa, of the good done by
them, not so much in directly converting the natives
to Christianity as in civilizing them, enlightening
their minds and raising their self respect ; and said,
that wherever the missionaries established a footing,
the arbitrary power of the chiefs declined. He
mentioned that in the territory of Port Natal, where,

1859. at the time it was first occupied by the English,
there were no natives,—all having fled from before
the Dutch Boers ; there is now a population of
120,000 natives, (I think this was the number he
stated) all fugitives from the tyranny of Panda and
other chiefs.   He spoke with just admiration of my
dear old friend Sir George Napier.

December 4th.

Drove out to Clapham to see my poor Aunt
Caroline ; found her in a very sad state of health,—
I fear, dying.

December 5th.

We went to Sandhurst, to stay with the William
Napiers, but as there was not room in their house
for both of us, Patrick Mac Dougall very kindly
lodged me in his house.   He and Captain Congreve
met us at dinner at William's.

December 6-8.

At Sandhurst, spending our time very pleasantly
with the William Napiers and their lovely children.
I have been very glad to have the opportunity of
improving my acquaintance with Mac Dougall, and
I find him very agreeable.   There is something
very attractive in his appearance and manners, very
winning, and I can readily believe what Emily says
of him, that he is a most loveable man.   At the
same time, handsome and agreeable as he is, neither
his face nor his conversation give me the impression of great intellectual power.   Very different

in this respect is Kingsley, with whom I have again
had some delightful conversation. He came over
to luncheon on Wednesday the 7th, and stayed two
hours, during which time we had some good talk,
though somewhat disturbed by other visitors. The
next day an exquisitely beautiful day, we drove over
to Eversley in Emily's carriage, and had a delightful visit to the Kingsleys. I am more and more
charmed with him in each successive visit. He is a
truly noble man. And the extent and variety of
his knowledge are astonishing. He is not only an
eloquent preacher and moralist, a poet and a
novelist, but an accomplished naturalist and antiquarian, an eager sportsman, what is he not? All
that he says bears the stamp at once of great
intellectual power and of a lofty and noble nature.
Unfortunately his health has suffered from too great
exertion of mind, and the physicians have ordered
him to write nothing for the next three years. We
talked of Froude's History, and he told us that he
had had thoughts of writing a historical novel on the
subject of the " Pilgrimage of Grace," with Robert
Aske for its hero. He talked much of Darwin's new
book on Species, expressing great admiration for it,
but saying that it was so startling that he had not
yet been able to make up his mind as to its sound-
ness. But were it merely as the result of thirty
years labour of such a man, he observed it ought to
be treated with reverence. He said that he had
himself been disposed to question the permanence
of species, and on the same ground to which Darwin
has attached so much importance, namely the

1859. great variations produced in the domestic races, but the very startling conclusions which Darwin has deduced from this doctrine have shown him the necessity of examining very carefully all the grounds of the reasoning. He is much interested also in the question of the flint " hatchets " or " arrow heads " in the drift. I told him of the curious fact mentioned at the last G. S. Meeting, of bronze ornaments discovered in the drift in Siberia. Kingsley remarked he had little doubt that gold had been worked in Northern and Central Asia, in very remote times, by nations probably now extinct, that Herodotus mentioned some of the Scythian tribes as possessing much gold, and that the very ancient myth of the Arimaspians probably related to the same facts. He told us much about the condition of Britain under the Romans, and about the Saxon conquest; thought that one principal reason why so little of Roman building remains *above ground* in this country, was, that the degenerate Romanized Britons were chiefly collected in walled towns ; and the Saxons, a wild race, who hated towns and town life, took these towns and destroyed them utterly. He doubts Pevensey being Anderida, as commonly supposed, for Anderida is described as situated in a forest, whereas Pevensey must evidently have been in those times, an island or peninsula in the sea.

The British Church, about which so much has been said, he suspects to be mythical, and doubts whether Christianity was at all generally or firmly established in Britain while it was a Roman province.

December 9th.   1859.

From Sandhurst to Brighton by Blackwater, Guildford and Reigate.

———

LETTERS.

Mildenhall,
March 9th, 1860.

1860.

My dear Leonora,

It has indeed been a severe and trying winter, or rather succession of winters—not over yet, for we have had heavy snow storms the last three days, and at this moment the ground is quite white.   There are hardly any flowers out yet, except Snowdrops and Winter Aconites, and here and there a few tufts of Crocuses. I find, on referring the register I keep of the flowering and leafing of garden plants, that vegetation is about as forward now as it was on the 14th of February last year. On the 28th of last month, the day after we came home, there was a furious gale of wind; by far the most destructive that we have had since Fanny and I have been living here. In these grounds, seven or eight trees were blown down or broken to pieces ; one fine old elm torn up by the roots, and another snapped off at twelve or fifteen feet from the ground.   In the plantations, a good many fir trees have been blown down ; in one place six large spruces are uprooted and lying side by side, with such masses of soil turned up by their roots, that they have the appearance of a barricade.   But the mischief here is trifling compared to what has

1860. been done in several other places in this district;
Mr. Newton is said to have lost a thousand trees,
and Mr. Waddington several hundred ; and in the
northern part of Norfolk, the destruction has been
enormous.   At Barton, I am glad to say, no
material damage has been done—I drove over to
Barton the day before yesterday, and thought
my Father looking decidedly better than when I last
saw him, in January—able also to speak apparently
with greater ease, and in better spirits.   Lady
Bunbury too was, for her, pretty well.   You will no
doubt, have heard that my Uncle Sir William
Napier died last month ; happily he died quite
tranquilly and without pain, which was a great
mercy, as he had suffered so terribly last winter.
He was a man of very great powers of mind and of
a lofty noble character.   Since the beginning of last
year the world has lost several great writers ; but
one does not, on public grounds, mourn so much for
those who, like Humboldt and Hallam and Sir
William Napier, had done their work and left great
and complete writings, as for Prescott and
Macaulay, cut off in the midst of their great and
admirable undertakings.

I read through Prescott's "Philip the Second"
while we were in London, and with very great
delight, and sorely did I regret that he had not been
permitted to live to finish it.   I am now **reading**
Motley's " Rise of the Dutch Republic :" **very**
interesting too, he is a very different writer from
Prescott, and less fascinating, yet he brings the
events and the characters of the time very forcibly

before us. Reading thus in succession two different
narratives of the same time, fixes the history more
forcibly in one's mind, and it is interesting to
compare the different views and feelings of the two
writers. Though both sympathize heartily with the
brave Netherlanders, and both have a hearty
detestation of the cruelty, bigotry and faithlessness
of their Spanish oppressors, yet they have different
predilections in various points. Egmont is evidently
Prescott's favourite; William of Orange is Motley's.
Reading these books gives me a desire to see the
towns of Holland and Belgium, and I hope we
shall some day make a tour in those countries.

Darwin's book has made a greater *sensation* than
any strictly scientific book that I remember. It is
wonderful how much it has been talked about by
unscientific people; talked about of course by many
who have not read it, and by some I suspect, who
have read without understanding it, for it is a very
hard book. Certainly it is a very remarkable work
of extraordinary power and ability, and founded on
a wonderful mass of careful observation, I confess
that, for my own part, though I have read it with
great care, I am not altogether convinced; possibly,
when I shall have seen the body of evidence which
he is to bring forward in his large work, I may
be better satisfied: but as yet, I doubt. It is,
however, a great triumph for Darwin that he has
made converts of the greatest geologist and the
greatest botanist of our time; at last, Joseph
Hooker so far adopts the Darwinian theory that he
considers it not as proved, but as a hypothesis,

1860. quite *as* admissible as the opposite one of permanent
species, and far more suggestive.   I wonder what
Humboldt would have thought of Darwin's book.

I am now reading Hooker's "Essay on the Flora
of Australia," exceedingly able and curious.   There
have been some valuable books of another kind
published this season. — Sir Emerson Tennant's
"Ceylon," which seems a most complete and careful
account of that beautiful island, in all its relations,
and very pleasantly written; I have read only the
natural history part, which is extremely well done.
Oliphant's account of Lord Elgin's "Mission to
China and Japan."   "MacClintock's Voyage" (this
I have not yet read) and "Russell's Diary in India,"
which interested me exceedingly.

Have Tennyson's "Idylls of the King" found
their way to Berlin?   I like them better, on the
whole, than anything else Tennyson has written,
though there is no part of them perhaps as powerful
as "Locksley Hall," nor any passage equal in
lyrical beauty to some of the songs in the "Princess."
One other book I must mention (though you will
think I am writing a library catalogue rather than a
letter), Kingsley's "Miscellanies:" some of the
Essays are to me delightful.   Indeed I delight in
Kingsley's conversation and in his writings.

I have not time to say much about politics;
indeed I might write much but say little or nothing
to the purpose, for I do not in the least see my way,
all appears cloudy, threatening and uncertain,—I
mean in foreign politics.

Within this tight little island all looks smooth

and prosperous enough at present, but there is no 1860.
knowing how soon we may be caught in the
hurricane of European war; therefore in spite of
Mr. Bright, I hold it very necessary to *keep our
powder dry.* I wish most heartily that all may go
well with the Italians; but I cannot help having
great fears.

Pray give my love to your husband,
Ever your affectionate Brother,
CHARLES J. F. BUNBURY.
(Sunday, March 11*).*

Mildenhall,
March 17th, 1860.

My Dear Katharine,

You were beginning to read Dr. Hooker's
" Essay on the Flora of Australia," when I was last
at your house, and I daresay you have long since
finished it. I have read it since we came home, and
with great admiration. It is masterly, like most of his
writings; and what an immense mass of well arranged
information it contains! What labour those comparative tables of species must have cost! I agree
with Susan that Joseph Hooker has real genius, and
he has besides—what does not always go along with
genius, but what is characteristic of the greatest men
in every department—a prodigious power of work.
His arguments in favour of the Darwinian theory
are in part not very clear to me; but if I understand
him rightly his advocacy of the theory goes no
further than this: that he considers it as a hypothesis, *as* admissible as that of separate creations

1860. (neither admitting of actual proof) and much more suggestive ; which I think may be admitted.

I very much approve of Hooker's phrase of "creation by variation," it appears to me to be an error to represent the question as between creation and the reverse ; it is really between two different modes of creation,—between sudden and gradual creation, we may say, or between intermittent and continuous creation. It is exceedingly curious what Hooker points out so clearly, the very great difference between the floras of south-western and south-eastern Australia, in the same latitudes ; also the large number of European plants, not introduced by man, that exist in the southern hemisphere. I have set to work again at the Indian fossil plants, my work at them having been interrupted last autumn by my Father's illness ; and I hope to get them off my hands in the course of the next two or three months. I have also a set of fossil Ferns to name for the Geological Survey. I have begun to look through and arrange the huge parcel of Indian dried plants that Dr. Hooker has given me :— a great acquisition to my collection. The number of species must be very large. The specimens are not in general very fine, or very carefully dried, (not like Mr. Fox's or Mr. Oakes's), and as far as I have gone yet, they do not give me the idea of a very splendid fllora—I mean, not of a tropical filora—however I have made but small progress as yet.

I am in the second volume of Motley's "Rise of the Dutch Republic ;" coming near to the execution of Egmont and Horn ; and exceedingly 1860. interesting it is ; painfully interesting. I think very highly of Motley as a historian : he tells his story very clearly, and succeeds in giving one an exceedingly distinct and vivid picture of the men and the times of which he treats ; and he appears to be very careful and painstaking. He is certainly a *good hater*, and does not disguise it, and assuredly there is abundant reason for it ; I do not think it would be easy to find a period of history, in which that disposition would be more called forth than it is by the Spanish government of the Netherlands. I never close my day's reading without a feeling of devout thankfulness that I did not live in those times. I am reading also, as a lighter book, the new series of " Friends in Council," and like it much.

And what are you doing ? I have no doubt you are abundantly and well employed ; but I have heard much less of you, since we have been in the country, than of the rest of the family. Fanny is pretty well, but she has been working—I will not say like a horse, but like a whole team—ever since we came home, and, as usual, rather overdoing it. I think your children would like to see her feeding her fowls ; I delight in them, they are so very tame and bold, crowding round her the minute she appears, and scrambling so for the corn. One very handsome cock in particular, by name Tommy, is gloriously impudent, jumping up and perching on her arm or her shoulder, to help himself out of the dish !

What a tangle and perplexity foreign affairs are

1860. in ! the people of Central Italy have behaved splendidly, and shown without any doubt or wavering what *their* purpose is. And then what will follow ? I trust the King of Sardinia will have firmness enough not to refuse the annexation ; and then will Austria attack him or wait to be attacked.

And the French Emperor, can he have the face to object to the annexation of Central Italy to Sardinia while he is clutching Savoy ? Those who have any faith at all in him must be more sanguine than I am.

I think the pretence that Piedmont may become so powerful as to be dangerous to France, is one of the most impudent I ever heard. There is an excellent article on Savoy in the *Daily News* of yesterday.

Pray tell Charles Lyell that Mrs. Abraham got a handsome ammonite out of the gravel pit on Barton hill ; that pit in which he and I have so often observed the contorted strata. Mrs. Abraham is away now in London, and the Evanses have been in London too, but I hope they came back yesterday.

We have had a cold spring hitherto, and our garden is very backward ; but this day has been beautiful ; multitudes of bees, busy about the Crocuses, water-wagtails running about the lawn, and the rooks busy and noisy about their nests.

Much love to Harry and your children, and to all the dear people at No. 17.

Ever your very affectionate Brother,
C. J. F. BUNBURY.

Mildenhall,
March 26th, 1860.

My Dear Susan,

I cannot resist the temptation of writing a few words on politics, suggested by the eloquent *tirade* in your last letter to Fanny. I must begin by saying that I most fully agree with you as to the abomination of this kidnapping of Savoy, and look upon it as one of the great political crimes (and they have not been few), of our time. But when you talk of our going to war to prevent it, I am a little puzzled as to the plan of campaign. At which of the seaports of Savoy would you land your forces? I understand there is not good anchorage at Chambery! Seriously our insular position, with its many advantages, has (it seems to me) this *disadvantage*, that we cannot give effectual assistance to any country that is not readily accessible by sea. Recollect what happened in 1805 : the battle of Trafalgar—the greatest naval victory that England ever gained, the greatest naval victory there has been since Lepanto, did not prevent Napoleon from marching to Vienna, nor from destroying the Austro-Russian army at Austerlitz. Again there is one peculiarly awkward point in this Savoy affair, as it appears to me, I cannot, for the life of me see how we could effectually resist the Emperor (by force I mean) without entangling ourselves in an Austrian and therefore an anti-Italian alliance. I confess that if I were an English Minister just now, I should feel myself almost paralyzed by this consideration, that a blow struck against Nap. in that

1860. part of Europe, would almost inevitably be a blow struck against Italy. If, indeed, the Piedmontese government were avowedly opposed to this French demand, and called on us for help, it would be a very different case; but I am afraid the Piedmontese are willing enough to let Savoy go. If Nap. were to try next to clutch Belgium, which he very likely may, our course would be much clearer. As for the Rhine—I take it that the fortresses of Cologne and Coblentz would be harder nuts to crack than even the famous Quadrilateral. These are my notions on the subject; I should be very glad to hear what you think of them. How admirably the Italians have behaved! I was rather amused at your wish to make a burnt offering of Lord Palmerston and the rest. I always thought you were somewhat of an incendiary, but to propose to burn a Temple* is going rather far. Fanny sends much love and thanks for your letter, and will write to you to-morrow.

We shall be delighted to see you here whenever you like to come.

With much love to all the family party, I am ever,

Your very affectionate Brother,

C. J. F. BUNBURY.

Mildenhall,

March 29th, 1860.

My dear Mrs. Horner,

I thank you and dear Susan very much for your kind and sympathizing letters. You are

---

* Lord Palmerston's surname.

always so kind and affectionate, and full of sym- 1860.
pathy, that I was sure you would feel for me
when you heard of my Aunt's death. I was indeed
fully prepared for it, and was not at all surprised,
for I felt sure when I last saw her, that her death
must be near at hand ; and it was happy for her,
that she followed her husband so soon and passed
away so quietly, without any further suffering or
lingering decay.

A better or a kinder woman never lived;
and as you say, her death has called up
into my memory many touching recollections of
the past, and of all her kindness to me from the
earliest times I can remember. I did not go up
to town to attend her funeral, because my Father is
in such a precarious and indeed alarming state,
that I was desirous to be in the way, whenever
he might wish to see me. The last account from
Mr. Image is merely that he is "no worse," which is
no great comfort at any rate ; but Lady Bunbury,
from whom I have a note this afternoon, evidently
thinks he *is* worse. It is a sad trial for her. I
cannot but be very uneasy about him.

I thank you very much for your kindness in
offering me a room at your house in case of my
coming up to town.

I cannot write comfortably on other subjects just
now, so will conclude with wishing you a pleasant
visit to Paris.

Pray give my love to all your family party,
and my best thanks to Susan and Katharine
for their kind letters ; I do not write to them

1860. because I could only repeat what I have now written to you.

Ever yours very affectionately,
CHARLES J. F. BUNBURY.

Barton,
April 16th, 1860.

My Dear Mr. Horner,

I have had no time for writing to you since the death of my dear and excellent Father. During the fortnight that his last illness continued there was hardly at any time a glimpse of hope of his recovery : and his sufferings were so severe, and his last struggles so much prolonged and so painful to witness, that we could not help feeling it quite a relief when the last moment came. It was a comfort that we were all around him when he died, and that only a few hours before, he knew us, and pressed our hands. Since he is gone, I feel, and shall continue to feel more and more the loss we have all sustained.

Yourself and Mr. Nicholl (my Father's London lawyer) are appointed trustees of the entail.

I am glad that he has thus testified to the just esteem and respect which I know he felt for you. The legacies are few, but in the number there is one of £100 to dear Susan, for whom I know he always felt a particular regard.

He has left everything in such perfect order and such clear notes and statements on every point, that Mr. Nicholl says there will be no difficulties at all. It is rarely that a country gentlemen has the 1860. good fortune that I thus have, of coming into a perfectly unincumbered estate.

But in one respect my Father was far happier than I, — that he came into possession of his property by the death of one whom he could not regret.

Poor Lady Bunbury has borne the shock hitherto, with more resignation and composure and less appearance of injury to her health than I could have expected, but I fear it will tell upon her more in a little while. Fanny is not well, I am sorry to say; her nerves and spirits have been sorely shaken by the terrible fortnight of anxiety. Since I began this letter I have received dear Mrs. Horner's kind note, for which I thank her heartily. I trust your visit to Paris has answered well, both for health and pleasure.

Believe me ever,
Your very affectionate Son-in-law,
C. J. F. BUNBURY.

Mildenhall,
April 22nd, 1860.

My Dear Katharine,

I thank you very heartily for your most kind, affectionate and sympathetic letter. Both this and your former one to Fanny went to our hearts; they came so evidently and truly from your heart, and bore such testimony to that affectionate and generous warmth of feeling which distinguishes you.

Friday was indeed a solemn and a trying day for

1860. all of us; yet it was far less painful, and shook me
less than the last two or three days of his life—those
days of struggle and suffering, when one was com-
pelled to pray that all might be over. I looked
several times on his face after the spirit had departed
—there was nothing painful or terrible *there*, all
was solemn, tranquil repose, a mild placid beauty;
all appearance of struggle and pain and anxiety was
over; he looked as if in a happy sleep, and the face
actually assumed the appearance of that of a much
younger man; you would have said he was not
above sixty. There is no pain but a kind of solemn
pleasure in remembering that look, far different
from the worn, harrassed look of physical suffering
in his last days. That was terrible to see and is
painful to recollect. I like much more to remember
him, and do constantly recall him to my mind's
eye, as standing before the fire in his own room,
or strolling with me round the arboretum and
pleasure ground looking at his favourite trees, and
enjoying the beauty of the grounds he had laid out.
Often I find it difficult to believe that what has
passed within these last few weeks is real; it seems
to me like a strange bewildering dream. It is only
by an effort of mind that I can convince myself that
I did not remain several months at Barton, instead
of little more than three weeks. As time rolls on, I
feel, and shall continue to feel our loss more and
more; it continually comes back upon me, that I
have no longer that wise and good Father to consult
or to apply to. It is as you truly say, a great
comfort to remember that he lived to so great an age,

prosperous and loved and honoured, and (to a very 1860.
late period) with so little of decay or suffering ; for
even last summer he was as fine and healthy an old
man as one often sees. It is a still greater happiness
that he died with all those he most loved around him,
conscious almost to the last of their affection,
knowing that he left them truly harmonious and
united among themselves, and that he would be
sincerely and universally lamented by all who had
known him. Truly he had done his work in this
world, and done it well, and is gone to receive his
reward.

I need not say to you, and indeed I cannot express
what a solace and comfort it has been and is to
me to have my darling Wife with me through all this
trial, and to feel that we are so entirely *one* in heart
and mind, in feeling and purpose. We are both
very happy to get back to our dear quiet old home,
and hope to have a few days of complete rest,
which we much want. Poor Lady Bunbury remains
at Barton for the present, but for *how long*, I believe
neither she nor anyone else can tell. She is very
very much to be pitied. It is a comfort to feel that
Fanny and I have done all that we could by consideration and affection to soothe her, and to
alleviate her sorrow ; and that she feels it.

I believe we shall have to go up to town on
business some time next month, and I shall then
hope to see you. I should like very much to ask
you to come here, but our whereabouts will for some
time be so much dependant on circumstances not
within my control, and my time will be so much

1860. occupied that I do not venture to propose anything. With much love to your family party,

I am ever,

Your very affectionate Brother,

CHARLES J. F. BUNBURY.

Barton,

April 29th, 1860.

My Dear Lyell,

I have too long delayed answering your letter; but during our few days of rest at Mildenhall I felt too lazy to write, and since we came back I have been too busy. I was much interested by what you told me in your letter, of the application made to you by the London University,* it was very honourable to them, and a very proper compliment to you, but at the same time I think you very wise in declining it.

My Father's mind continued clear almost to the last, and happily he was able to use his eyes in reading to a very late time. On the Easter Monday, only four days before he died, after we had been talking in the morning of Sicilian affairs, he sent for me again to show me the Athenæum newspaper which he had been reading, and to point out to me particularly an article on Owen's Palæontology. These were almost the last distinct and intelligible words that I heard him speak. Only a very few days before, he had marked a long passage in Irvine's Sicily, describing the eruption

* Charles Lyell was asked to stand for M. P. to the London University, which he declined, as he did other honourable offers, determining to keep to his one great pursuit.

of Etna in 1809, and caused his amanuensis to 1860.
copy it out, and charged me to give it to you, as an
independent corroboration of the account he had
given you of that eruption. I will bring the paper
when we come to town. He has left some very
interesting notes—unfortunately too brief, and I am
afraid incomplete, of what he had seen in Sicily and
the Lipari islands; he intended to print them
together with some other papers for private cir-
culation.

Fanny is rather better and sleeps better. I hope
you and dear Mary are well. We have at last
beautiful bright weather, and much milder these last
two days. This place is looking beautiful, but there
is a sadness over its beauty.

With much love to Mary,

I am ever,

Very affectionately yours,

CHARLES J. F. BUNBURY.

———

Barton,
May 13th, 1860.

My Dear Katharine,

I take advantage of a Sunday, and an
excessively wet Sunday, to write to you, for you may
suppose I have been tolerably busy since we
returned hither.

I am going on steadily with the work of gradually
making myself acquainted with the concerns of the
estates which have come into my possession. Here
at Barton, everything seems clear and in the
most perfect order, and I have the advantage of an

1860. invaluable assistant in Scott, who has long enjoyed my dear Father's utmost confidence, and is thoroughly conversant with all his plans and views and system of management ; he is a pattern of order and punctuality, and one of the most clear-headed men I know. As far as I yet see there is little left for me to do *here*, but to keep up what my Father has established. At Mildenhall, apparently, there will be a great deal more to be done, and it is a much more troublesome parish. We are now placed in a position of great trust and responsibility,— one in which we may have the means of doing much good, and in which much will be required of us : and I feel it will be no easy matter to keep up to the example which my Father has left me. At the same time I should be very sorry to neglect intellectual improvement, or to let my mind be entirely absorbed by the cares of business, and I hope still to find time occasionally for botany and geology. I have lately resumed my reading of Motley, and have just finished the second volume. How intensely interesting is that history of the struggles of the Dutch against their oppressors. I do not know anything in history more exciting— more glorious—more sublime I may say, than the defence of Haarlem and Leyden. Certainly there is nothing more heroic in Greek or Roman history. I have lately got Owen's new work on " Palaeontology," but as far as I have yet looked through it, it seems rather a useful dry book of reference than anything to read.

This place is looking beautiful, and daily more so,

as daily something new bursts into leaf or flower. 1860.
How I should delight to shew it all to you. But we
shall probably see you in London before long, at
least I hope so.

With love to your husband and children,

I am ever,

Your very affectionate Brother,

CHARLES J. F. BUNBURY.

Barton,
June 12th, 1860.

My Dear Susan,

Fanny will be very happy to give her
name and £5 towards the fund you mention for the
relief of the wounded and widows of Garibaldi's
expedition. I think it an admirable scheme, and
one of the best possible ways of showing one's
sympathy. What glorious and astonishing success
he has had! It does one's heart good to read of
what he and the Sicilians have done. He is a true
hero; and I do not say this because he is successful, but because his success has been gained by such
a union of daring enterprise and sagacious com-
binations. I trust it will now be but a little while
before Sicily will be thoroughly free, and for ever,
from the Bourbon dynasty and its soldiers. Your
letter to Fanny interested and entertained us both
very much; but I am sorry to see that you have a
leaning towards that dangerous scoundrel, the
Emperor of the French—however, I have not time
at present to launch into a political disquisition.

We are deluged with rain here, and are longing

1860. almost desparingly for fine weather.   The farmers
are very uneasy.   Three trees in our grounds here
—two of them in the pleasure ground—were struck
by lightning last Wednesday.

With much love to all the family party,

I am ever your very affectionate Brother,

CHARLES J. F. BUNBURY.

———————

Barton,
June 13th, 1860.

My Dear Lyell,

I must give you some account of what
happened here this day last week, the 6th.   I meant
to have written to you sooner, but have had no time.
On that day early in the afternoon, we had a short
but severe thunderstorm; one flash, a little before
two, was most startlingly vivid, and was followed with
hardly any interval by a violent peal of thunder.   I
found afterwards that three trees in our grounds had
been struck—a Silver Fir tree and an Oak beside
the pond at the corner of the pleasure ground, and
an Elm in what we call the Vicarage grove.   The
Oak which is a tall young tree, I suppose about 34
years old, presents the most remarkable appearance.
A strip or ribbon of bark has been torn clean off, in
a somewhat spiral line, from near the top of the
tree quite down to the ground, at least twenty feet;
and where this stripped line touches the ground, the
soil is excavated in a little pit, and the mould and
loose flints scattered about as if by a minature mine.
To some distance on each side of the wound (as I
call the line on which the bark had been stripped

off) the bark is completely loosened from the 1860. wood, so that I can easily slip my hand between them ; and along the middle of the wound the wood also is ploughed up and torn into long slender splinters. But no part either of the wood or the bark shows the slightest trace of burning or scorching ; nor does the herbage at the root of the tree appear in the least singed. The Silver Fir which stands about ten feet from the Oak, has no continuous wound down its trunk, but large pieces of bark, at various heights, have been torn off and scattered about as if blown off by gunpowder, and other portions of the bark are loosened.

There are two other trees, a Chesnut and a Beech, which stands close to these two, on either side, actually touching them with their branches, and neither of these are affected in the slightest degree.

The third tree which was struck, a large old Elm, stands in a grove several hundred yards to the S. E. of the other two trees : and this also has had a narrow strip of bark torn off it, reaching down its stem from a great height to the ground ; the whole thickness of the rugged old bark, with a sturdy stem of Ivy that adhered to it, being torn off, and the wood laid bare along that single line. The appearances altogether, especially in the case of the Oak, seem to me very much in favour of the hypothesis (Dr. Wallaston's was it not ?) of the sudden conversion of the sap into steam, and consequent explosion ; but it is curious that the action should have been so much limited to particular lines.

1860.   We have had very extraordinary weather ever
since we returned from town, an alternation of
storms of wind with rain, and storms of wind without
rain : the most tempestuous and inclement June, I
almost ever remember. The blossoms and young
foliage of our trees are wofully disfigured by the
furious gales : and the young tender shoots of the
Conifers have been shrivelled as if by a severe frost.
The farmers are seriously uneasy.

Much love to dear Mary,

Ever yours affectionately,

CHARLES J. F. BUNBURY.

Barton,

June 24th, 1860.

My Dear Katharine,

What glorious weather for Ferns. If we
could but transform ourselves into Ferns or
Epiphytes how we should thrive ! At least—I do
not know how it may be in London, but *here* we
have been living for some days past in a perfect
vapour bath,—a warm air as well, saturated with
moisture as in the rainy season within the tropics :
a Devonshire climate, very fine for vegetation
generally, and especially (as I said before) for
Ferns : but more relaxing and enervating than I can
describe.

Fortunately there was a fine day for the Flower
Show last Friday ; it was at Ickworth ; I met there
a good many people whom I knew, and some whom
I was very glad to see ; but it was by no means so

great a gathering as here at Barton last year. We
are going to Mildenhall on Tuesday next, to stay
there a week, and then probably to Lowestoft for
three or four days before we return home.

Poor Lady Bunbury intends to leave this place
to-morrow on her long journey into Wales ; it will
certainly be a great effort for her—a great wrench—
to remove from here, but I am convinced it will be
best.

I look forward with real delight to seeing you
here in the course of the summer, and I shall have
various things to show Leonard that I think will
please him. Franky, I think, will approve of the
swing, which we are going to have put up.

I have no time for study except in the evenings,
but I have at last finished "Motley," which I think
you have begun,—and very interesting the concluding scene is. I dare say you will be disgusted
as I was, with the style of the introductory chapter,
but it is well worth while to read on ; the second
and third volumes are far better than the first. I
have now begun the fifth volume of " Froude."

I hope it is true—as Edward writes, that he hears
it reported—that Louis Napoleon has returned from
Baden without having gained anything by his journey.
I wonder whether he means to try to annex Ireland ;
it seems there are some pamphlets newly come out
at Paris which point that way.

With much love to all your belongings,

I am ever your very affectionate Brother,

CHARLES J. F. BUNBURY.

Barton,
October 3rd, 1860.

My Dear Emily,

I knew that you would hear from Cissy all
about us, as well as about her and Henry. It is a
great pleasure to us to have them and the dear
children in the house. And just now we have a
good house-full of children, for Katharine and
Henry Lyell are here with their four: and the two
sets of *juveniles* are great friends.

The first two days that Cecilia was here, we were
away on a visit to the Thornhills at Riddlesworth.

The worst result of this cold and wet season has
been the total failure of the potatoes, the disease
among them being worse (at least in this part of
England) than in any year since '45. Here, at
Barton, the failure of the potato crop may be said
to be *total*. Poor John Brand told me some time
ago, that he had had to dig up thirty yards of
potato ground, before he could get potatoes enough
for one meal. At Mildenhall, things are not quite so
bad, but yet worse than ever before. This failure,
and the very high prices of food, will make the
coming winter, I fear, a very hard and trying one
for the poor, and they will require much attention
and assistance. And, if White of Selborne's rule be
correct, we have reason to apprehend a hard
winter in every sense:—he says he had observed,
that a severe winter, with much frost and snow,
was generally preceded by a very wet and chilly
summer.

The trees, many of them at least, are changing

colour this year very early. The Beeches in the 1860.
tall grove on the North of the house are already
richly variegated. The best Scarlet Oak (near the
library windows) is almost entirely red, though far
from being so vivid as in the last two or three years:
the Norway Maple has almost everywhere changed
colour, and the Virginia Creeper is in great beauty.
The common Walnut, which has been very poor in
foliage all this year, is already quite leafless, but
the Hickories have not changed colour. There are
some cones on the great Cephalonian Fir, but so
high up that one cannot see them distinctly. The
fine Cryptomeria in the arboretum, which was
loaded with cones the last two years, has scarcely
one this year. The Catalpa has blossomed tolerably
well, but shows no disposition to form pods.

I do not well remember what was the date of my
last letter, but I rather think it is since then that I
have finished the 6th volume of " Froude's History,"
ending with the death of Mary. I scarcely ever
read any thing more interesting or more touching
than his narratives of the deaths of the Protestant
martyrs in Mary's reign. I have since begun "Lord
Mahon," and am nearly through the first volume,—
very pleasant, easy reading, not at all profound. I
am rather diverted by the political *soreness* which
peeps out continually in this first volume, when the
author was fresh from the struggle of the Reform
Bill. Fanny sends her love to you,

Believe me ever,

Your very affectionate Step-son,

CHARLES J. F. BUNBURY.

1860.   Barton,
                                          November 8th, 1860.

My Dear Mr. Horner,

I have this morning attended the funeral of
my excellent old friend Mrs. Sparkes—certainly,
beyond my own nearest relations, one of the oldest
friends that remained to me, for our acquaintance
began when I was only nine years old. She was an
excellent woman, and most warmly and constantly
attached to me ; I feel her loss much ; but it is
a comfort that she died amongst us, and knowing
how much she was valued.

I quite agree with you in heartily approving of
Lord John's letter, and thinking it a sound and
manly exposition of true Whig principles—the best
principles of the best times of Whiggery.   It cannot
fail I think, to make a great sensation on the
Continent, and to strengthen the influence of
England there.

I think foreign affairs look a little less threatening
—less like general confusion—than they did some
little time ago ; it would appear as if the Warsaw
Conferences were not likely to lead to any immediate
practical result, but it is impossible to feel quite
easy so long as the chief ruler of the most military
and most ambitious nation in the world continues
to pursue a policy so utterly dark, devious and incomprehensible ; if indeed it be a policy at all, and
not a mere aimless succession of caprices.   It seems
we are now to believe that the monstrous interference of the French fleet at Gaeta was solely with
a view to the personal safety of the ex-king of

Naples, and to ensure his being able to run away 1860. in safety! As for Rome, it is likely to be in a pleasant condition between the French garrison, the people (not the mildest or least fiery in the world) and the rabble of adventurers from all countries that the Pope is recruiting for his own army. Truly, if there be any English travellers going thither this winter, I think they must be bold people.

It is curious how soon poor old Admiral Napier has followed Lord Dundonald with whom he had a certain analogy in his character and actions.

We have had for some time past, very fine weather for the time of year; we have still some Fuchsias and Geraniums in blossom out of doors, untouched by frost, and the conservatory is very gay with flowers. I have had several trees taken down, and the Laurels generally near the house reduced to moderate dimensions.

I wish you a very successful geological campaign, and hope that I may be able to take part in some of the actions.

Fanny sends her love.

With much love to Mrs. Horner and all your party, I am ever

Yours very affectionately,

CHARLES J. F. BUNBURY.

———

Barton,
November 11th, 1860.

My Dear Katharine,

I am sure you will have been sorry to hear of the death of our dear good old Mrs. Sparkes, my

1860. friend of forty-two years. I was very much pleased with a sentiment I met with in one of Leslie's letters (in his Autobiography) where he says he is convinced that death is a calamity only to the survivors, for " I am sure that God takes us all whenever it is best for us."

By the way, have you read that book " Leslie's Autobiography ? " Leslie, the painter. It has interested and pleased me very much. I remember I once sat next to Leslie at a dinner at the Moore's, and thought him very agreeable, and this book has given me so pleasant an impression, that I regret I did not know more of him.

I am now reading at odd times a novel which Susan recommended, " Madamoiselle Mori." It is very pretty.

I see by the papers that Mr. Kingsley was justified in his doctrine of the healthiness of this wet season, for it appears by the official returns, that this last quarter has been one of the healthiest on record. This agrees with what I heard from Mr. Lovelock as to Mildenhall. Touching my own individual personality, I must confess that a hotter and dryer season is apt to give me a more decided feeling of positive and vigorous health ; but I have nothing to complain of.

The last time we were at Mildenhall, about a fortnight ago, we found everything about the garden and house in excellent order, and in particular I was very agreeably surprised with the flourishing and vigorous condition of the Ferns in our little bit of a greenhouse. There were Onychium lucidum, Woodwardia radicans, Aspidium falcatum, 1860. three or four Adiantums, Asplenium bulbiferum and Asplenium flaccidum, all in splendid condition; the Asplenium flaccidum, grown so luxuriant, as to be hardly recognisable, with pendulous fronds, two feet and a half long, and bearing quantities of little baby plants on its pinnules, just like bulbiferum. I hope we shall multiply it in this way, and then I will keep a plant for you; in the meantime if you would like to have a frond, I am drying some. I have got a new part of "The Second Century of Ferns," chiefly Aspleniums. I have also got the new numbers of the Geological Society's Journal, in which I find much to read; and am diligently studying a long German paper on the fossil plants of the brown coal, which Charles Lyell lent me.

The dressing bell has rung, so I must leave off with much love to all your party.

Ever your very affectionate Brother,
CHARLES J. F. BUNBURY.

Barton,
November 23rd, 1860.

My Dear Susan,

I wish to put up in our church here a monumental tablet to my dear Father's memory, and I should be very much pleased if you would give me a design for it. Of course, I do not wish you to trouble yourself about this, if your time or thoughts are much taken up by other objects, but if you can take this trouble, I am sure your taste

1860. and feeling would devise something much more pleasing to me than what any one else would be likely to supply : and I think it will be an additional motive for you to comply with my petition, that he had a great regard for you.   You understand that it is to be a mural tablet and not of very large size, or it would be out of proportion with the church.

I hope dear Mrs. Horner's sciatica is relieved,—I was very sorry to hear of her sufferings.

I do not think the late accounts from Italy are comfortable.   The continual resistance which the ex-king, encouraged by the grand Juggler of France, as well as by the Pope, is keeping up at Gaeta, evidently tend to promote and inspirit reactionary movements, and to keep up the hopes of those who look to a counter-revolution ; and I am afraid it is true that there exists among the people in power a very bad and unworthy feeling towards Garibaldi and his followers.   What a glorious man he is !   It is a pretty sketch of his troops in '49 in the 2nd volume of "Mademoiselle Mori," a book which I have read on your recommendation, and with great pleasure.

Addio, dear Susan.   Much love to all the family party.

Ever your very affectionate Brother,
CHARLES J. F. BUNBURY.

Mildenhall,   1860.

November 30th, 1860.

My Dear Joanna,

It will be a great pleasure to us both to see you at Barton next week. Dear little Arthur* too we shall be most happy to see. Your dog† shall also be welcome.

The Mac Murdos have been with us ten days: he is very agreeable besides being a man of very remarkable ability, and very remarkable character.

Pray thank Susan for me very much for her kind and agreeable letter, and for her promise to comply with my request. Perhaps you will not mind looking into Barton church while you are with us, and then you will be able to give her more full information than I could do. I hope to see you soon. I will write no more just now. With much love to all the family party,

I am ever your affectionate Brother,

C. J. F. BUNBURY.

Barton,

December 29th, 1860.

My Dear Emily,

I was very sorry indeed to hear that you had met with such a severe and disagreeable fall; but I trust you will not have been seriously the worse for it, and that you will soon and thoroughly get over its effects. I had intended, before I heard

* Arthur Lyell.

† Thistle, her Scotch terrier, who passed his last days at Barton, and is buried there.

1860. of this accident, to write to you, on the close of
this year, which has been marked by a loss so heavy
to all of us,—to you so irreparable. There is
always, at least when one has passed the middle of
life, something solemn in looking back on the
year that is closing; but how much more when the
event that stands prominently out to our memory,
is the loss of one, on whose like we shall not look
again. Be assured, I am not likely to forget him.
I hope I shall always look on the example of his
virtues as the noblest part of the inheritance he has
left me: and my ambition will be so to conduct
what depends on me, that if he could return, he
would have no reason to be dissatisfied with what
had been done since his departure. I will say no
more on this subject, except to wish you health and
all the happiness that is possible, in the year
which will soon begin.

We have such severe weather as I have seldom
felt; quite an old-fashioned Christmas time, I
fancy: bright and clear and sunny, with intense
frost, the surface of the snow quite crisp. In the
night between Christmas eve and Christmas day,
the thermometer in the garden was down at 1 deg.
below zero of fahrenheit, or 33 deg. below the
freezing point. Last night it was at 9 deg., and this
day felt the coldest of all, because there was more
wind. I am afraid you must be very cold in
Wales. I feel nervous about the evergreens, and
others of the more tender trees and shrubs, and
the more so, as this severity of frost came rather
suddenly, after very mild weather in the early part

of the month, so that the Fuchsias and Hydrangeas 1860.
and the deciduous Cypress had not yet lost their
leaves when the frost came, and the dark China rose
against the west front of the house was still in
full blossom, and the Banksian roses in leaf.

(December 30th). A furious snow storm last
night : thaw and rain to-day. I have a lot of
feathered pensioners, who come every morning for
the crumbs I throw out on the gravel walk before my
window.

I must tell you how much I like and admire
Montague Mac Murdo. They stayed with us ten
days, so that I had the opportunity of much conversation with him ; and I was exceedingly struck
with his uncommon vigour and power of mind,—
with the range and grasp of his intellect, as well as
the force of his character. I shall be much disappointed if he does not show himself a really great
man. He is moreover a remarkably pleasant one.
Poor Susan was far from well during the time she
was here. More lately, we have had Mr. and Mrs.
Kingsley with two of their children, staying with us
for four days; and thus having so good an
opportunity of improving my acquaintance with Mr.
Kingsley, I am most entirely confirmed in the high
opinion I had formed, as well of his agreeableness
as of his higher qualities, intellectual and moral.
He is a man thoroughly after my own heart. I like
Mrs. Kingsley also very much, and the girl and
boy are particularly pleasing young people.

This past year has been, to be sure, well filled
with stirring and extraordinary public events, and

1860. the next does not promise to be quiet or uneventful either. There are plenty of unsettled questions, both in the Old World and in the New; plenty of trains laid for new struggles :—Rome, Venice, Hungary, Holstein, and the two great *sections* of the United States. As for this last queston, I only hope that the Northern people will be firm, and as they have at last taken courage to bring in a candidate of their own choice, that they will stand by the principles which he is supposed to represent and not yield like cravens to the bluster of the Southerners.

Edward means to set out about the 17th of January for Rome, Athens and Constantinople. I do not mention the good people at the Cottage, as I know you hear of them regularly. I hope you will come here and stay some time with us before we go to London, which we talk of doing about the 12th of February.

Believe me ever your affectionate Step-son,

CHARLES J. F. BUNBURY.

———

Barton,
December 30th, 1860.

My Dear Susan,

I have no time for more than a few lines to thank you very heartily for your letter, and for your design, which I think extremely pretty and very suitable to its purpose. If I can get it well executed I am sure it will be a great ornament to the Church. I admire your design much and thank you very much for the trouble you have taken.

I condole with you on the severity of the season, 1860. as I think you are no fonder of cold weather than I am ; still it is better than a winter of continual rain and gloom. I most heartily wish you and all the dear people at 15, 17 and 53, a happy new year, and many of them, with health and every blessing.

Ever your very affectionate Brother,
CHARLES J. F. BUNBURY.

My Dear Susan,

I must add a scrap to Fanny's letter to say how entirely I agree with your remarks on Sir William Napier's History.* The main faults of the book (and they run throughout it) are that his disgust at the blunders, follies and cruelties of the Spaniards, have made him too insensible to the noble spirit with which they defended their independence, and on the other hand, his enthusiastic admiration of the military and political genius of Napoleon has blinded him to his wickedness. These are great faults and yet it is a noble work, and I am very glad you are reading it. For my part I am deep in Mackintosh's " Ethical Philosophy."

Your very affectionate Brother,
C. J. F. BUNBURY.

---

* " History of the Peninsula War."

Barton,
                                             January 1st, 1861.

My Dear Mrs. Horner,

Many thanks for your very kind note, and every good wish for the new year ; may it bring health and happiness to you and all who are dear to you. I am very sorry that you still continue to be so much tormented by that painful and harrassing complaint of neuralgia, and that the doctors do not seem able to remove it : it must be sadly wearing. I am a good deal worried by an obstinate cold and cough which hang about me, accompanied by a great deal of depression and languor : a sort of influenza, I suppose. But indeed the weather is very trying, and the new year has not opened auspiciously in that respect. Fanny I am happy to say, is just now pretty well, and has been very busy all to-day promoting the New Year's festivities of the young people : for we have a large party of juveniles in the house, Louisa, Cecil and Clement, who arrived yesterday, and little Emily and Henry : and they have been very merry.

Joanna's visit here was an exceedingly great pleasure to me, and I was heartily sorry to part with her. I felt when she was going away, that I had not talked half as much with her as I wished.

I am indeed much grieved at the news of Mrs. Richard Napier's alarming illness : very alarming, I fear, considering her age and her previous delicate state. Poor Richard Napier, I feel very much for him.

With much love to Mr. Horner and all your 1861.
family party, I am ever,

        Most affectionately yours,

            CHARLES J. F. BUNBURY.

                      Barton,
                 January 2nd, 1861

My Dear Joanna,

      I am very glad to hear that you mean to
make an addition to your cabinet of shells, and I
was much pleased also to hear of Leonard's studies
and acquisitions. He is a very promising young
naturalist indeed, and I hope to live to see him
President either of the Linnean or Geological
Society. Shaftesbury's " Characteristics " I never
read, and know them only by a pleasing account of
the book and its author in Sir James Mackintosh's
" Discourse on Ethical Philosophy." Gray on the
other hand, in his letters, speaks contemptuously of
the work ; but it is quite possible he may have been
prejudiced.

Fanny has been to Bury to-day, and seen the
Louis Mallets, who are coming to stay with us from
Monday to Wednesday next.

Cecil and Clement have been indulging in the
luxury of sliding on a pond. The two little ones,
Emmy and Henry had a capital feast and game of
romps with their cousins here yesterday, and were
in high glee.

I see the Emperor of Austria has thought it
advisable to make a show of magnanimity in the
case of Count Teleki,—influenced probably by the

1861. expressed opinion of Louis Nap. We were very happy to hear to-day from Mary that Mrs. Richard Napier was better.

With much love to all the family, I am ever,

Your very affectionate Brother,

C. J. F. BUNBURY.

Barton,

February 5th, 1861

My Dear Mr. Horner,

I thank you very heartily for your truly kind and cordial letter, which I received this morning. The very warm feeling towards me which you express so kindly, and which I know you feel is a source of true gratification to me, and I trust I may always retain your good opinion; as I am sure I may safely say that I shall always feel a most warm and hearty attachment to you and yours.

I am happy to say that I have at last packed up the greater part of the Nagpore fossil plants, and dispatched them in two boxes bound for the Geological Society ; the greater part, not the whole, for I have kept a certain number in hopes of being able in the course of this year to give them a further examination, and also some very bulky pieces which I could not conveniently pack just now. The specimens I have sent off are all labelled to the best of my power, and I will try hard to get my notes put into shape so that I may be able to read them at any meeting that may be most convenient for the purpose during my stay in town.

I wish you joy of having finished your Presidential

Address, which must indeed be rather a formidable 1861. task, and which will, I have no doubt be less vague and more instructive than the French Emperor's Address to the Chambers. I hope to be in London in time to hear you deliver it, and also to dine with the Society afterwards. At any rate I shall be much obliged if you will secure for me a place at the dinner table. I thank you much for your gift of Hugh Miller's book, though it has not yet arrived. I read the Cruise of "The Betsy" while at your house, and was much pleased with it, but I have it not in my collection, and it will be very welcome.

Pray give Mrs. Horner my hearty thanks for her very kind letter and for Leigh Hunt's pretty book which has just arrived.

With much love to all your family party, believe me always,

Your affectionate Son-in-law,
C. J. F. Bunbury.

Barton,
February 5th, 1861.

My dear Mary,

I thank you very much for the pretty little book of Mrs. Jameson, which you so kindly sent me on my birthday, with your kind good wishes, for which I am equally grateful. I shall, I have no doubt, have great pleasure in reading these essays of Mrs. Jameson's: in everything she wrote there was sure to be something worth attending to,—fine thoughts and noble sentiments, just remarks happily expressed, and a fine appreciation of what was

1861. beautiful and exalted in art and nature. She was a
great loss. It is astonishing, when one comes to
reckon them up, how many eminent and admirable
men and women have been taken from the world in
the last two years.

We returned yesterday from Mildenhall, having
spent a pleasant and busy fortnight there, and had
very fine and pleasant weather almost the whole
time : but now it has changed—the weather I mean,
and is cloudy, blowy, cold and disagreeable. Fanny
declares that the sun shines only at Mildenhall ! but
this I do not admit.

This morning I have been taking a survey of the
killed and wounded in the garden, arboretum, and
pleasure grounds here, and they are dismally
numerous, and some that will not easily be replaced. I lately finished Lord Stanhope's ("Lord
Mahon's) History of England," very pleasant reading, and since then I have read one volume of
"Lord Malmsbury's Correspondence," which does
not give one a high or favourable idea of Courts.
He was for several years ambassador at St.
Petersburg, while the great Catherine was on the
throne, and at a very eventful time, while we were
engaged in the American War ; and to be sure, his
letters do not inspire us with much veneration for
her, her favourites, her ministers, or anything that
was hers. He had before been for some years at
the Prussian Court, and there are some curious
particulars about Frederick the Great in his latter
days, and his nephew and heir. I am anxious to
read the new volumes of "Motley," but I like, as

far as I can, to preserve a certain degree of sequence 1861.
and connexion in my historical reading.

The world is just now in a promising state for
making history for future writers to treat of. East,
west, north and south, what a complication of actual
or impending agitations and wars; and the
Americans who have so long looked on in quiet at
the tumults of Europe, seem now likely to have
their full share. And we ourselves, though hitherto
peaceful at home, cannot help feeling a nervous
thrill through every fibre of our cotton system, when
we think of the possible effects of a civil war in the
States. It is in every way an anxious time.

I hope we shall be in town by the 14th, and I
look forward with great pleasure to seeing much of
you and Lyell during our stay there.

With much love to him,

I am ever your very affectionate Brother,
CHARLES J. F. BUNBURY.

Barton,
Feb. 6th, 1861.

My Dear Katharine,

Very many thanks for your kind letter and
good wishes. I am happy to say our tender Ferns
in the houses were perfectly preserved from all the
severe weather, and look now as well and flourishing
as ever. It was a great pleasure to see them in
such beauty, when I first went into the greenhouse
after coming from Mildenhall. Out of doors too,
the Hartstongue and the common Lastreas look as
green and bright as possible. But the destruction

1861. of other things has been lamentable. All our fine
standard Roses which poor Lady Bunbury took so
much pride in—all utterly killed ; nearly all the
climbing Roses the same. The gardener says we
have lost at least fifty distinct varieties of Roses.
All the Bays and Laurustinuses, the Magnolia
grandiflora and Cratægus glabra, killed. Of the
Conifers, our fine Cephalonian Fir, I am glad to say
appears quite unhurt; so also the Cedar of Lebanon,
the Pinus excelsa, Abies Smithiana, the Cryptomeria
and most of our Deodars, but we have lost Pinus
insignis, Cupressus funebris and several others, and
even the Araucaria has all its young branches killed.
But till the Spring is fairly set in, we shall not know
really what mischief is done. There has been no
such winter certainly within my recollection. It was
curious to see at Mildenhall, that the snow was
hardly gone before the ground under our trees was
perfectly spangled with the little yellow Aconites ;
and Snowdrops were out before the first of this
month. We spent a pleasant and busy fortnight
there, but my time was so much occupied with
packing up two collections of fossil plants and
completing my notes on them, that I had not much
time to attend to other departments of my museum.
Lady Cullum has just been here, and she and I
have exchanged condolences about our Roses.

With my love to your Husband and children,
believe me ever

Your very affectionate Brother.

C. B.

JOURNAL.

February 20th.

We yesterday came up to London, in the same 186 1. railway carriage with Henry and Cissy, and are established at Lillyman's Hotel, in Brook Street.

February 24th.

Read a good deal of the "Autobiography" of Mrs. Piozzi (Johnson's Mrs. Thrale), edited by Mr. Hayward: very entertaining. It is not properly an Autobiography, however: the greater part of the first volume is filled with an introductory memoir by the editor; and the rest made up of the lady's letters, and her copious notes on various books relating to Johnson and herself.

We dined with the Charles Lyells : no one else of the party except Mr. Horner. Charles Lyell and I had a good deal of talk about America. We afterwards spent half-an-hour with Lady Bunbury, who has been so ill, that this is the first time I have been allowed to see her since we came to town. She seems very infirm, but talked well and with animation. Speaking of Lord Malmsbury and his negotiations in Holland, she told me that when she was in that country with Lord Clancarty (in 1814) she had heard and read much of the Revolution of 1787 ;—that the Stadtholder of that time (who was afterwards an exile in England) was an utter nonentity ; but his wife (Lord Malmsbury's friend)

1861. an exceedingly clever woman, and of great force of character. Lady Bunbury described her as having singularly penetrating and expressive eyes. She had been hated by the people of Holland in 1787, but when after the war, her son was restored and made King of the Netherlands, she had the wisdom to abstain entirely from any meddling in State affairs, and to live a quiet, private and useful life.

———

February 25th.

We went in the evening to the Meeting of the Geographical Society. A prodigious crowd. We got tolerable places for hearing, but not for seeing the maps and illustrations. A most curious and entertaining lecture (for so it was, rather than a paper) from an extraordinary man, a M. Du Chaillu, an Americanized Frenchman who had lived eight years in the midst of the forests of the wildest, hottest and wettest part of Western tropical Africa. He spoke broken English, and with a very strong foreign accent, but with infinite spirit, clearness and vivacity.

His description of the habits and manners of the Gorrilla and Chimpanzee, and of another great ape of the same genus inhabiting that country, was exceedingly animated and entertaining. I hope it will all be in print, so I need not do more than put down a few hints to help my memory. The Gorilla a most extraordinary and tremendous animal; all that this traveller learned fully confirms the impression given by the information that Owen had collected. M. Du Chaillu penetrated into the very heart of the Gorilla country, killed no less than twenty-one Gorillas, and has brought their skins and skeletons to this country. His description of the tremendous strength and fierceness of the Gorilla, of the horrible roar he utters when angry, and which can be heard three miles off: the noise (like that of a big drum) which he makes by beating his hands on his chest; the effect of these awful sounds in the midst of those dark and gloomy forests, was amazingly striking. Besides the Gorilla and Chimpanzee, there is another species of Troglodytes in those countries, which this traveller has named Troglodytes calvus, and of which he gave a very curious account : especially of the ingenuity with which it twines together branches and twigs of trees into a roof or shed to shelter its family. Also of the comical cunning and thievish tricks of a young one of the kind which he kept tame.

The Gorilla he found to be quite untameable, however young.

The Secretary of the Society, Francis Galton, then spoke of the physical geography of that part of Africa ; and afterwards Owen gave us some very interesting remarks, in his own admirable way, of the various resemblances and differences of the Gorilla as compared with man. Among other things he pointed out that the proportions of the chest and shoulders, and arms and all the upper parts of the body,—the bones and muscles of these parts—are larger and more powerful than in any human being, considerably larger than those of the Irish Giant,

1861.   whose skeleton is in the museum of the College of
Surgeons ; but the legs are those of a puny dwarf in
proportion ; so that the height of the animal when
set upright is only about 5ft. 6in., though if it
had the human proportions it would be at least
8ft. high.   Again, with such astonishing physical
development, the brain is not larger than that of
a human infant of six months old.

March 1st.

March has not begun in orthodox style, but with
violent showers and bright gleams of warm weather
between.

March 2nd.

Went with Fanny to the Panorama of Messina,*
with which we were both delighted.   That of Rome
we did not like so well.

Went to the Museum of Practical Geology ; saw
Huxley, who told me that he had examined some
reptilian remains in sand stones sent home from the
Damoodah or Burdwan beds of Bengal, and had
ascertained them to belong to the genus Dicynodon.
This is likely to be an important geological fact, as
that very remarkable and characteristic reptile the
Dicynodon has hitherto been found only in South
Africa, in beds of which the geological age is thought
to be pretty well ascertained.

Salter showed me some very fine specimens of the
Adiantites Hibernicus in the yellowish sandstone,

Messina in 1809.

Upper Devonian (?) from Ireland; some of them 1861.
apparently in fructification.

March 3rd.

Went with a family party of Horners and Lyells
to the Zoological Gardens, and spent some time
there. Saw the two Yaks *(Bos grunniens)*, rarities
newly arrived here; very remarkable looking animals
with their bodies clothed with a thick coat of long
hair almost reaching to the ground, and their magnificent bushy white tails.

Saw also that most extraordinary and monstrous
looking bird the *Balæniceps*, or "whale-headed
Crane," from the White Nile; the appearance of
its head and beak is like some wild grotesque
caricature; one can hardly look at it without
laughing. The Dodo itself could not have been
more uncouth. Another curiosity is the *Bateleur*
Eagle from South Africa,—so named by Le Vaillant
—remarkable for the beauty of his colouring and
his odd and seemingly affected gestures.

Also the great Bats, called "Flying Foxes"
(Pteropus, the Roussette of Buffon), of which there
are now three or four, kept in a cage in one of
the warmest houses; very singular beasts, somewhat
foxy in colour, but darker than the common Fox;
the head not so much pointed as that of a Fox, but
tending that way. Their attitudes and movements
are extremely curious to watch.

March 5th.

At the Athenæum, met Louis Mallet :—very full

1861. of Prince Napoleon's great speech in the French Senate, which is indeed very remarkable, both for the power and ability it shows, and for the line of policy it points out. It is so long since there has been room for anything like oratory in France, that this has the greater effect. The fact of the Emperor sending specially to Prince Napoleon to congratulate him on this speech, shews that the sentiments of it are at least not offensive to him ; and it seems to promise well, both for Italy and for the duration of peace between France and England.

March 7th.

First sitting to Mr. Boxall for my picture.

Linnean Society's meeting.—First, a paper by Bentham, of remarks on various points relating to the Menispermaceæ, Bixaceæ, Samydaceæ, and some other tropical families of plants.

Next, a very interesting and curious paper by Joseph Hooker, on the Vegetation of Clarence Peak in the Island of Fernando Po.

Clarence Peak is a mountain 10,000 feet high, and has been ascended by Mr. Mann : — the first mountain, in Western Tropical Africa, of which the summit has been reached by a European. Hooker gave us a lively description of the difficulties of the ascent. The flora of the upper or temperate region of the mountain, though not rich in species, is very remarkable from its very close agreement both in genera and species, with the flora of the mountains of Abyssinia at so great a distance. It also

includes many European genera, and even some 1861.
well-known European species, but has scarcely any
thing in common with the vegetation of the Cape.

After this had been read, Mr. Bell gave a short
explanation of M. Du Chaillu's discoveries and
views concerning the physical geography of the
interior of tropical Africa, with reference to the
probability of a chain of mountains or great high-
land, connecting the mountains of Abyssinia with
those of the coast opposite to Fernando Po. I
made some remarks on the peculiarly insulated
character of the Cape flora, which were confirmed
by Bentham from his examination of collections
from the Natal country. Hooker spoke of the
strong South European element in the vegetation of
the Abyssinian highlands : said that in the collections
at Kew from that country he had found 200
European species.

———

March 9th.

Went with Fanny to Henderson's nursery garden,
St. John's Wood, near the Regents Park. The
most remarkable things we saw here were, Lapageria
rosea, in flower, a fine climber, like a Smilax in
habit, with firm dark green leaves, and large
beautiful rose-coloured blossoms, pendulous and
bell-shaped. Rhododendron jasminiflorum, one of
the Sikkim species (?) with beautiful delicate white
flowers, with remarkably long slender tubes, looking
at first sight quite unlike Rhododendron flowers.

Afterwards went with dear Katharine to the

1861. Botanic Gardens in the Regents Park. The large conservatory here in great beauty: full of Azaleas and other ornamental plants. Fine trees of Rhododendron arboreum, in full blossom, with their splendid heads of deep scarlet flowers. Grevillea robusta has grown to a great height, and is a remarkably beautiful and graceful tree, with its feathery drooping foliage, very unlike the usual stiff hard character of the Proteaceous trees, but it does not flower. We noticed also Araucaria Bidwillii, excelsa and Cunninghamii : the very curious and eccentric Phyllocladus ; a Casuarina, and several interesting Ferns.

March 10th.

Read a good bit of the Autobiography of Dr. Alexander Carlyle, lately published, edited by Mr. Burton ; his account of the Jacobite rising in 1745, and the battle of Preston Pans, of which he was an eye witness, very entertaining. He was one of the unlucky Edinburgh volunteers of that time. His narrative quite agrees in the main with those which we had before, but only makes it more and more wonderful how the Royalist army could contrive to lose the battle. Carlyle slept in a cottage close to the lines of the Royal army : he was wakened by the first cannon that was fired, and he says it was not above fifteen minutes at most before he sallied out, but already the fields were covered with the scattered fugitives of Cope's army, and with the pursuing Highlanders. It was, I suppose, one of the *shortest* battles on record. There was probably disaffection as well as cowardice. Poor Colonel 1861. Gardiner told Carlyle two days before the battle, that he did not think that ten men of his regiment would follow him in charging the enemy.

We dined with Harry and Katharine Lyell ; met among others, Mr. Edgeworth ; a distinguished Indian botanist, half brother to Miss Edgworth, Sir Edward Ryan, Mrs. Agnew, (a lady who has lived many years in India, and who obtained for Katharine that splendid set of Assam Ferns, of which I have had so liberal a share) and Huxley the great Zoologist. In the evening some very agreeable women : Mrs. Galton, Mrs. Phillimore, Miss Moore, besides the sisterhood.

Spent an hour in the botanical room of the British Museum, examining the very interesting collection of stems, fruits, and specimens of vegetable structure.

News came yesterday of the surrender of the Citadel Messina, so that the whole of Sicily is now in the power of the Italian national party. This is good.

News of the Duchess of Kent's death.

Lyell tells me that that caricature of a bird— the Balæniceps, which I saw in the Zoological Gardens, on the 3rd of this month, has died.

March 20th.

Geological Society's Meeting. My paper on
the "Fossil Plants from Nagpore," (which has hung
on my hands two years !) was read. Then Huxley
gave us, not a paper, but a *vivâ voce* account, in his
clear and spirited manner, of the remains of the
Dicynodon, and of a fish called Ceratodus, which
he had determined in specimens from the Burdwan
coal formation ; and he showed the inference to be,
(though not with certainty) that the formation is
Triassic.

_____

March 21st.

Linnean Society meeting. Flowers were exhibited
—produced in this country, I forget in what garden,
of the curious Megacarpæa polyandra, the only
known Cruciferous plant with indefinite stamens,
in fact almost the only known Crucifer that
is decidedly anomalous in its structure.

The specimen exhibited had something in its
appearance that reminded me both of a Cauliflower
and of Thalictrum flavum, but it might also be
compared to the flower of the Horse-radish.
The stamens, which are much the most conspicuous
part of the flower, vary in number in different
flowers, but are always more than six.

I had a good deal of talk with Dr. Hooker, Mr.
Bentham, Mr. Ball, and the President Mr. Bell.—
The President is now the possessor of the house and
estate at Selbourne formerly belonging to Gilbert
White, and he tells me that he has the books and
bookcases and many other relics of the great
naturalist.

I spent some time in the Zoological Gardens very pleasantly with Joanna and Leonard. The day was extremely fine and mild—like summer, and the gardens appeared to great advantage. We noticed the Sloth, which is kept in one of the warmest houses, and was this day in a remarkable lively state, travelling from perch to perch about his den, in his curious fashion, with uncommon alacrity, always hanging by his claws, with his back downwards, exactly as described by Waterton, and in this manner moving from perch to perch, or along any one perch, with greater speed than one would have thought possible. The Roussette Bats, or Flying Foxes, in the same house, are not less peculiar in their movements. In another very warm house (the same in which are the Fennec Foxes and the Cercoleptes) there is another very rare and curious animal, the Douroucouli, that nocturnal monkey discovered by Humboldt and Bonpland. It sleeps during the day, and looks like a little ball of fur, with only its long and bushy tail extended.

The Yaks showed themselves to great advantage to-day, and are certainly very remarkable looking animals. The birds (my favourite class of animals) and especially the water-birds, were in a state of great animation on account of the fine spring weather. The Peacocks and many different species of beautiful Pheasants, made a glorious show. We noticed also particularly some beautiful Pigeons; the richly coloured purple Water-hen from Egypt, the "Manchourian" Cranes from Japan; very

1861. stately and handsome birds; the Saras Crane from India, still more remarkable for height; the Secretary bird and Cariama (or rather the Seriema?) of Brazil, with its curious mixture of characters of the Waders and of the Gallinaceous birds.

Leonard Lyell, who is only 10 years old, has a wonderfully exact and complete knowledge of the animals in these gardens, so that no one can be a better guide to them. His knowledge of natural history generally is very remarkable for his age, and the activity of his mind and his eagerness for information most striking.

March 26th.

We dined with Sir John Boileau : a pleasant party. Among the company was Mr. Elliot,* who was lately Secretary of Legation at Naples, with his wife. I had a good deal of talk with Mr. Elliot.

March 27th.

We travelled down to Barton entirely in the old way, by posting along the old road with which I used in former days to be so familiar, and made out our journey very pleasantly and comfortably.

* Afterwards Sir Henry Elliot.

LETTERS

Barton,
April 1st, 1861.

My Dear Katharine,

You have heard from Fanny how success- 1861.
fully we made out our old-fashioned journey from
London. It was really very comfortable and
pleasant : only the last stage or two began to be a
little wearisome. The country on the other side of
Chesterford is prettier than I remembered. It is a
great loss in travelling by railway along that line,
that one neither passes through Epping Forest nor
within sight of Audley End. We found everything
here in most admirable order, only the quantity of
dead evergreens, which had not yet been cut away,
gave rather a melancholy look to the grounds.
To-day the gardener has begun clearing them away.
We can now judge rather more exactly than before,
of the amount of mischief done by the winter,
though there are still several things which are in a
doubtful state, and may or may not recover. In
addition to the losses I was aware of before, the
common Italian Cypresses seem to be all dead or
dying, and they are a great loss. In fact, all the
species and varieties of the genus Cupressus that I
had here, appear to be killed : while of the numerous
species and varieties of Juniperus, not one appears
to be hurt. It is very odd. Of the Roses, the

1861. destruction turns out not to be quite so complete as
we had supposed ; a good many of the standard
ones, Pettit* finds, are killed only a certain way
down, and retain life in the lower parts of their
shoots : and he thinks they will in the end do well.
The conservatory is in great beauty, with a variety
of Azaleas and other handsome things : and in the
grounds there are plenty of pretty spring flowers.
In the Shrub wood the ground is perfectly carpeted
with wood Anemones and Primroses. The Adoxa
is nearly in blossom, and I will not neglect to dry
specimens for you. It is a very neat delicate little
thing. I find that Bentham does refer it to the
Caprifoliaceæ,—I cannot imagine why : it looks to
me as unlike that family as anything can well be.

Fanny is pretty well. Poor Cissy has been very
unwell. Her children in high health and spirits,
very saucy and funny, and very pretty. Henry
tolerably well. We had reckoned on seeing the
Kingsleys this afternoon, but have had a telegraphic
message to say that they cannot come till late
this evening, which makes me rather uneasy ; I
hope Mrs. Kingsley is not going to disappoint us
after all.

With much love to your husband and children
and all the family circle,

I am ever,

Your very affectionate Brother,

C. J. F. BUNBURY.

---

* The Gardener.

1861.

Barton,
Sunday, April 7th, 1861.

My Dear Mr. Horner,

Many thanks for sending me Mr. Nicholl's letter, which I return to you. It is all right. I talked the matter over with Mr. Nicholl while in London. It would take more time than I can spare to explain what the New England Company is ; but almost the whole parish of Eriswell belongs to them, and they are owners also of a little bit of fen land in Mildenhall. This little bit of land was subject to the payment of tithes to the Rector and Vicar of Mildenhall ; the Company therefore (like most of the fen proprietors) agreed to buy off the rectorial tithes ; the affair was in progress, but not completed when my Father died ; it therefore had to be done over again, and the sale to be effected in the names of the Trustees of the Entail, *i.e.*, yourself and Mr. Nicholl. I have indeed been very much grieved and shocked and so has Fanny, by the news of poor Henslow. He is a very great loss.

I hope dear Susan got home comfortably. Her visit was a very great pleasure to us both, and I was very sorry to part with her. I hope Mrs. Horner is better.

I have read your Presidential Address with great interest ; and Mr. Kingsley to whom I lent it, has also been much struck with it. With much love to all the family circle, I am ever,

Your very affectionate Son-in-law,
CHARLES J. F. BUNBURY.

I have made an exception in your favour to my general rule of not answering any *business* letter on a Sunday.

4 S 2

JOURNAL.

1861. These ten days, during which I have enjoyed
Kingsley's society and conversation, have been very
delightful to me. Every fresh opportunity of cultivating his society, adds to my regard and admiration
for him ; and I flatter myself that the liking is
reciprocal. I have seldom met with a finer or more
cultivated mind : at once refined and vigorous, though
he has clear, strong and decided opinions, he has
much toleration for variety of opinion : more indeed
than I should have inferred from some of his writings.
But I think his mind has been much ripened and
sobered since the days of "Alton Locke" and of his
earlier sermons. His conversation is peculiarly rich,
various, and instructive in matter, with great power
of expression. There is an especial charm to me in
good intellectual talk, which is not dogmatical, nor
excessive and overwhelming. Kingsley has a
particular delight in talking of the natural sciences,
and much of our conversation turned on those
subjects. He indeed told me that natural history
was his favourite study, and that to which he would
have devoted himself by choice if he had been
entirely at liberty to follow his own inclinations and
had no duties to draw him in other directions.

We spent two mornings here (at Mildenhall) in
looking over parts of my herbarium, which he
seemed to enjoy very much. But he has also great
knowledge and taste in the fine arts :—Susan Horner

says more than most people of her acquaintance : 1861.
and his reading appears to be very extensive in most
departments.

Our party was altogether very agreeably composed ; the Kingsleys, Minnie, Sarah, Susan, and
Mr. Smith of Combhurst,—all delightful in their
various ways.

Barton, April 24th.

Yesterday we two, with Minnie and Sarah, went
over to Cambridge, and spent a most agreeable day
there with the Kingsleys. Starting by the early
train, we arrived at their house in Fitzwilliam
street in time for breakfast ; and we returned to
Bury by the seven o'clock train. After breakfast
we went to the Fitzwilliam Museum, where we
first saw the pictures ; of these there is an account
in Waagen's book, but besides these mentioned by
him, we noticed particularly a Guido—" Herodias's
daughter with the head of John the Baptist."

Afterwards we went, Fanny, Minnie and I with
Kingsley, into the Library of the same Museum,
where he introduced us to the Librarian, and we
spent a long time looking over most curious books
of prints.

The collection of engravings is immense, and
they are admirably mounted and arranged. We
looked through the very fine collection of Marc
Antonios. Kingsley showed me the prints of
Phillipe Galle, and of the Sadelers (John Raphael
and Egidius), artists of whom I knew little or
nothing, pointing out in a most interesting manner,

1861. their characteristics, their merits, and defects.   He
told me that these artists were especially patronized
and employed by the Jesuits, and contributed
important help to the great Anti-protestant re-action
of the lattter half of the 16th and beginning of the
17th century, by the multitude of devotional prints
(according to the Jesuit teaching of religion) which
they executed and disseminated.   Accordingly a
large part of their works are of this nature : series
of martyrdoms, illustrations of lives of saints and
hermits, and a variety of subjects connected with
" Mariolatry."   Among others a very numerous and
curious set of prints of the " History of Ignatius
Loyola ;" a very curious set of " Triumphs of the
Cross," being a collection of martyrdoms of all ages
and countries in which the Cross was used ; another
of hermits, and another of Virgin Martyrs.

Also series of the Bible History, in which there is
a great deal of beauty as well as singularity in the
treatment of some of the subjects ; others of the
Passion of Our Lord ;   and   some   mythological
subjects.   Kingsley remarked that these artists
succeeded much better in small figures, and
especially in subjects including a multitude of such,
than in large single figures.   He said that after
them was a great decline in the art of engraving,
and that it was first revived in later times by the
French.

After luncheon we all went (a large party) to dear
old Trinity College ; admired the noble great Court,
the Hall, Neville's Court, and the Library ; looked
at the manuscripts of Milton, Newton and Byron

in the Library, and admired the statue of Byron.
Then Fanny and Mrs. Kingsley went back to rest
awhile, and the rest of us walked through St. John's
and through the delightful old walks (so familiar to
me) at the back of the Colleges to King's College
Chapel. Here the ladies stayed to hear the service.
Kingsley and I walked to the Botanic Garden and
spent a considerable time there with much enjoyment. The situation is very good, and the ground
apparently well laid out ; the out-door department
indeed looked at present rather dismal, from the
very evident ravages of the late terrible winter; but
we spent most of our time in the houses. I was
surprised to find the collection of hot-house and
greenhouse plants so rich and interesting, as I had
understood that the Cambridge garden was deficient
in these departments. The intelligent gardener
who attended us, indeed, said that there was a want
of funds for the purchase of new plants, and that
many of those we noticed were very old. The collection is of course not to be compared with those of
Kew or Paris, nor is it by any means equal to that
of Edinburgh, nor (at least in number of species) to
that of Berlin ; but it contains many interesting
things. Many beautiful and curious Ferns—among
them Polypodium vaccinifolium (erroneously labelled
lycopodioides), and Psilotum triquetrum, which I
remember was quite new to me when I first saw it
in the old Botanic Garden here, in 1831. Some
good Orchids and Begonias, few Palms. Passiflora
racemosa in full blossom, fine ; Passiflora alata in
flower. But some of the most interesting things are

1861. in the cooler houses.   A beautiful Diosma, of the
Adenandra group (Adenandra speciosa?) covered
with blossom.   The exquisite little Gentiana verna
in full beauty.   The delicate pretty little Houstonia
cærulea (which I never before saw alive) also
flowering abundantly.   Cheiranthus mutabilis in
flower.   The Mastic Shrub looking healthy, though
not large.

   Kingsley, besides the regular public lectures which
he gives as Professor of Modern History, is engaged also to give a private course of historical
lectures in his own house, to the Prince of Wales.
These are on Modern English History, — the
Hanoverian period : while his public course relates
to the middle ages.   We were allowed to hear
(through the door, being in the adjoining room) one
of these *esoteric* lectures, in which he gave a very
clear and very striking narrative of the battle of
Fontenoy.

   Kingsley is much with the Prince of Wales, and
will I hope and trust have great influence with him ;
I am sure his influence, as far as it goes, must be
altogether good.

   While we were at the Botanic Garden, Fanny
and Minnie and Mrs. Kingsley went to see
Sedgwick, who I am sorry to hear, is in a very
infirm state of health.   He gave them a most
interesting account of poor Henslow, whom he had
lately visited at Hitcham.   Henslow is, it seems,
inevitably dying : the doctors give no hope of his
life, though he has lingered much longer than had
been expected : but his intellect is perfectly clear,

and he is in the most cheerful and delightful frame 1861.
of mind, not merely resigned but happy to die.
The description of his state, which I heard at
second-hand from Fanny, strongly reminds me of
my dear Father's state in October, 1859.

I have read little this month, principally because
we have had so many friends staying with us,
and partly also because the weather which has been
generally fine, or at least dry, and the lengthening
days are unfavourable to indoor occupations. I
have, however, gone very carefully through Les-
quereux's account of the Coal Formation of
Pennsylvania (in Rogers's Geology of that State) an
important scientific work, from which I am making
copious notes.

I have read the 5th volume of Macaulay's
"History" which indeed I had begun in London. It
is a fragment only, but a magnificent fragment.
The period comprised in it, from the Peace of
Ryswick to the death of William the Third, is not
particularly interesting in itself : at least, not fertile
in great events or great characters, though it was
the prelude to great things ; but Macaulay has
succeeded in giving to it a remarkable degree of
interest. There is not the least sympton of decay
in this fragment ; all his characteristic merits and
faults are in full blossom. The narrative of the
death of William is perhaps equal to anything he
ever wrote.

1861.   Most striking too is the picture of the decrepit
state of the Spanish monarchy, under its last
Austrian King, Charles the Second.   The relation
of Portland's embassy to Paris, and that of Czar
Peter's visit to England, are also interesting.   I
have heard the account of the Darien scheme complained of as tedious, but I thought it far otherwise.
Altogether this volume revives and increases one's
regret that we can have no more from this delightful
writer.   Of the famous "Essays and Reviews," I
have as yet read only two : Dr. Temple on "The
Education of the World," and Goodwin on "Mosaic
Cosmogony."   I am at a loss to find anything
shocking or offensive, irreligious or unchristian in
either of them.   Goodwin's doctrine is nothing else
than what Charles Lyell has long maintained,—that
those who have striven to vindicate the Mosaic
Cosmogony by making out that the discoveries of
science are not at variance with it, have treated the
Bible narrative with no great reverence, by twisting
and explaining away its plain meaning, while they
have distorted the facts of science, and after all have
produced but an imperfect agreement.   Obviously
the only resource is to believe that the inspiration
or divine aid which was given to the Hebrew law-
giver, and the rest of the sacred writers for special
purposes, did not extend to questions which can be
investigated by man's ordinary faculties.

Dr. Temple's essay is very interesting, and
appears to me to be written in a truly reverential
and religious spirit.   I do not mean that I entirely
go along with him ; some parts of his reasoning

appear obscure, and others far-fetched and over 1861.
ingenious, but it is quite incomprehensible to me
what any reasonable man can find in it to offend
him.

LETTERS.

Barton,
April 30th, 1861.

My Dear Joanna,

I know that Fanny has written full
accounts of our proceedings, and especially of our
delightful day with the Kingsleys at Cambridge. It
was indeed most interesting. The only thing
Fanny could not give you an account of, was the
Botanic Garden, which Kingsley and I visited alone.
I mean without the rest of the party. I was
surprised to find so many interesting plants under
glass, as poor Henslow had told me that the
collection was not good except in the out-door
department. I wonder very much *whom* they will
find to succeed him in the Professorship.

We have sadly wintry weather—bitterly cold :
but the birds do not seem to mind it, they are
singing gloriously : I have not, however yet seen any
swallows. Sand martins indeed we saw at Mildenhall as early as the 10th ; but neither the chimney
swallow nor the house martin have made appearance : at least I have not yet seen them. The
cuckoo we have heard every day since Tuesday
the 27th.

Lady Napier (Minnie's Mother) is with us now as
well as dear Minnie and Sarah : but I am afraid

1861. they must leave us soon ; there will be great sorrow at parting. I hope Mrs. Horner is better, and that if we ever do get any real warm summer weather she will be quite set up. I am tolerably busy, have just gone through the Barton estate accounts for the last quarter, and have plenty of work still before me besides more agreeable studies, for I take business easily *and do not wear myself out.*

With much love to Mr. and Mrs. Horner and the Lyells, I am ever,

Your very affectionate Brother,

C. J. F. BUNBURY.

P.S.—A very agreeable letter the other day from Edward,—charmed with Athens.

Since I wrote the body of this letter, I have seen several swallows.

JOURNAL

May 8th.

The weather for the last five or six days has been cold enough for the depth of winter.

The recent news from the (no longer United) States of America has been very serious, and I cannot help now believing (which I did not at first) that both parties are really in earnest, and mean mischief. For a long time the seceding states on the one hand, and the Federal Government on the other, confined themselves to wordy manifestoes and "demonstrations," which I thought meant nothing but bluster. But at length the South struck the

first blow, by bombarding and taking Fort Sumter ; 1861.
and since then the President's appeal to the North
to arm, an appeal which seems to have been
warmly answered, the attack by the Virginians on
the Government Arsenal at Harper's Ferry, and
its destruction by the garrison ; the sanguinary
tumult at Baltimore, and the destruction of the
navy yard and shipping at Portsmouth, in Virginia,
to prevent their falling into the enemy's hands—
have followed in rapid succession. The civil war
has really begun, and it is likely enough to be a
savage one. The position of the President and
legislation at Washington—enclosed on all sides by
hostile or quasi-hostile States—is a dangerous one,
and there can be little doubt that the first serious
operations of the war will be in that quarter, as it
will be the great object of the Southern States to
gain possession of the Capital, and of the Northern,
to keep it. In a very short time probably we shall
have news of severe fighting at or near Washington.

_____

May 15th.

According to the recent news from America, the
aspect of affairs there seems to be considerably
changed. The Southerners have missed their
opportunity; there is now so great a force of
Northern troops collected at Washington that the
place is not merely considered secure, but there is a
talk of their taking the initiative, and beginning
operations against Virginia. Things certainly look
now much more favourable to the Federal cause,

1861. than they did a week or two ago.   This day I received and began to read Du Chaillu's "Travels in Equatorial Africa."

_____

Heard of the death of poor Henslow, who has at last been released from his sufferings, after lingering much longer than had been expected.   He is a very great loss, to his parish especially, one that can hardly be repaired.   He was a most excellent and valuable man.   The amount of good that he did by his energetic, enlightened, clear-sighted benevolence and his active, practical character, was inestimable. He might probably have been a very eminent scientific man, if he had thought fit, but for many years past his time and energies have been devoted much more to educational and practical than to scientific objects.

It is many years since he has brought forward anything in the nature of original scientific research ; indeed I am not aware of anything since the paper on a "Hybrid Digitalis," read before the Cambridge Philosophical Society in 1831.   He was a capital lecturer.

Henslow's age was only 64, and he had all the appearance of a strong man ; so that one's regret for his death is heightened by a feeling that it was premature.

_____

Yesterday I read a very interesting and well-written article in the *Edinburgh Review*, on that

admirable man Alexis de Torcqueville. He was a
great political philosopher: a truly wise, enlightened
and thoughtful patriot : as amiable in his private,
as wise and honest in his public life. The observations on the nature of democracy, on the English
character, on our Indian Empire, as quoted in
this Review from his letters are admirable. How
deplorable that so wise and good a man should have
been in any way mixed up with the French expedition to Rome,—one of the worst political crimes
of our time.

The very fine and warm weather this week has
brought vegetation forward in a remarkable manner.
The rapidity with which the leaves and flowers have
come out is very striking, and the colouring of
spring is now fast ripening into that of summer.
The wild spring flowers seem uncommonly abundant and fine this year, and I have never been more
struck with their beauty.

———

May 28th.

Finished Du Chaillu's "Adventures in Equatorial
Africa": a very entertaining book. The style is very
lively—not by any means correctly English, being
full of Americanisms; but free from that offensive
flippancy and affectation of *slang*, which are so
frequent in the books of sporting travellers. The
country which he explored,—within a few degrees
N. and S. of the equator, on the W. coast of Africa
—from the Moondah River on the N. to Cape Lopez,
and the Nazareth River on the S.—was almost

1861. unknown to Europeans, and he has given us some curious information, though less I confess than I had expected from the way in which he was cried up by the Geographical Society. Of botany and geology he evidently knows nothing, and from some indications I am inclined to suspect that his zoological science is not very profound. His gorilla hunts are very well told; the characteristics and habits of those monstrous apes, the impression produced by them, and his various encounters with them, are described with great spirit and force, and are extremely entertaining. The same may be said of his accounts of the curious nest-building Ape—the Troglodytes calvus—whether it be a variety of the Chimpanzee or a distinct species; of his history of the terrible Ants called Bashikouay; and of the notices of the habits of some other wild animals.

There are also some curious notices of the native tribes: in particular of their horrible and atrocious superstitions respecting witchcraft; superstitions which seem to transform even the mildest of them into raging demons for the time. The strangest of all his discoveries, and to me, by far the hardest to believe, is that of a nation of habitual *cannibals* :— cannibals in a more intense degree than any of the Tribes of Brazil or Polynesia, for they actually deal in human flesh, habitually buying and selling in their markets !

I shall be inclined to class Du Chaillu, as a traveller, in the same category with Major Harris, rather than with Park or Livingstone or Denham, or any of the higher order of travellers.

May 30th. 1861.

The seventeenth anniversary of our happy—truly happy—wedding.

May the Giver of all good accept my humble and grateful thanks for permitting us to enjoy this return of the day in health and peace, and perfect harmony and union of hearts and minds.

June 2nd.

Lyell writes to me,— "I wish time may do as "much for Du Chaillu as he has in verifying Abyss- "inian Bruce and old Herodotus, but after talking "with the 'Missing Link,' as they call this modern "intermediate, between the Gorilla and Cuvier, "I have had no small misgivings,—even before "Gray's attack. That the 'native products' in which, "according to Murchison, the young Gorilla hunter's "father dealt, were negro slaves, few seem to doubt. "That the son made excursions into the interior, "and collected most of his specimens himself, I "do not question ; but he did not inspire me with "much confidence ; and a good anatomist, in a "letter to Dr. Boot, says that 'Du Chaillu, when "shown skulls of the Gorilla and the Chimpanzee, "had no idea which was which."

=====

LETTERS.

Mildenhall,
June 7th, 1861.

My Dear Mr. Horner,

I have been exceedingly shocked to-day by hearing of the death of Cavour. It is a dreadful

4 T

1861. misfortune, and I cannot at all calculate what may or may not be the consequences. There were few men of our time, I take it, on whose single life so much depended.

I have finished the Spanish chapter of Buckle's second volume, which is very interesting, and I am just coming to the Reformation in Scotland. He is quite in his element in treating of Spain ; it is an instance peculiarly strong, and a striking illustration of the effects of superstition and (what he is pleased to call) loyalty. I think, as far as I yet judge, that this volume is better than the rest, though I do not expect it will be as popular.

Pray give my best thanks to Susan for her very agreeable letter, with much love to her and to all the family party, and believe me

Ever your affectionate son-in-law,

CHARLES J. F. BUNBURY.

Mildenhall,
June 9th, 1861.

My Dear Katharine,

I am writing to you in my dear old Museum here, which, however, is about to be shorn of some of its glories : for I have made arrangements for removing part of the dried plants, and hope that the boxes containing my Ferns will make their journey to Barton to-morrow ; I move them first, because I have already there a very large set of Indian Ferns which Dr. Hooker has sent me, and which I cannot put in order till my arranged collection is in the same place. The living Ferns

here, both in the greenhouse and in the open air are
flourishing; Asplenium bulbiferum and flaccidum,
Woodwardia radicans and Cyrtomium falcatum in
great beauty; the out-door kinds in a beautiful state,
with part of their fronds just newly unfolded, and
part in the *crosier* state. At Barton, besides the
Ferns you saw last year, we have young plants of
Cyrtomium caryotidum, Cheilanthes farinosa, and
Pellæa flexuosa (alias Pteris, alias Platyloma, alias
two or three other names), all from Henderson's,
very flourishing, and growing fast. I was
pleased to find Lady Anne Lloyd* interested about
Ferns, and we compared notes, and she admired the
growth of ours. I gave a lot of dried specimens
of Ferns to Rose Kingsley, who is a charming and
most intelligent girl. I spent some hours delightfully, the last time we were here, in looking over
parts of my herbarium with Kingsley, who seemed
very much interested by it: for, in addition to all
his other studies and accomplishments, he is a
zealous botanist; indeed he has a passion for all
branches of natural history, and has repeatedly said
to me that it would be his favourite occupation if
he could afford to devote himself to it.

I hope Leonard is as zealous a naturalist as ever.
Did he gain any acquisitions in that way while
abroad? Did he see the Jardin des Plantes?

Our old evergreen Oak (Ilex) in the paddock here,
which I believe must be above a century old,
really appears to be killed by the cold of this last
winter; it is not as far as I can discover, shewing

* Wife of Lady Bunbury's Cousin.

1861. the least sign of life anywhere; while the Ilexes at Barton (which also had had all their leaves killed) are sprouting out again vigorously. The great Walnut tree in our kitchen garden here, likewise looks very much as if it were dead, and indeed the Walnut trees everywhere about this part of the country are either killed or very much hurt. This is not the case with the American Black Walnut or with the Hickories—at least at Barton. A fine Holly hedge at the Nursery* is killed to the ground.

Pray give my love to Harry and your children, and believe me,

Your very affectionate Brother.

CHARLES J. F. BUNBURY.

JOURNAL.

June 12th.

It does not seem to be doubted that Cavour's valuable life was sacrificed through the barbaric ignorance and unskilfulness of his medical attendants.

June 19th.

We came to London and established ourselves as before at Lillyman's Hotel. The weather very sultry this afternoon; it has been beautiful for the last eight or nine days.

I attended the Council of the Geological Society

* At Mildenhall.

and dined at the Club where there was a good 1861.
muster. I sat between Sir Roderick Murchison and
Huxley ; had some talk with the former on African
discovery, and much with the latter on various
topics.

Huxley is an exceedingly clever man, and rather
an agreeable one : and I believe a good man, but I
cannot take very cordially to one so entirely without
veneration.

At the evening meeting of the Geological Society
(the concluding one of the session) several short
papers were read ; the most interesting to me was a
notice of the breaking out of a new Volcano on
the African coast of the Red Sea, not far from the
Straits.

June 20th.

Went to the Meeting of the Linnean Society, the
last of the season, Bentham in the chair. I exhibited a flowering specimen of the Indian Horse-
chesnut, Æsculus (or Pavia) Indica, which has just
come into blossom at Barton, and gave some
account of it. There was a fine specimen exhibited
of a very curious sponge, Hyalonema I think is the
name, with a beautiful long plumelike tuft of glassy
silicious filaments, like a feather of glass : it had
been sent from Japan by Mr. Veitch. Huxley gave
an account of it, and explained its structure, about
which there has been much doubt ; he believes it
to be a true sponge, and the wonderfully glassy
filaments to be an exaggeration (as it were) of the
silicious spicula, common in sponges

1861.   Hooker read a capital paper on the Oaks of
Syria and Asia Minor, part of the results of his last
year's tour.   He treated of the three most common
Oaks of those countries,—Quercus pseudococcifera,
Ægilops and infectoria, showed their great variable-
ness, and how great a number of false species have
been made out of them.   Quercus pseudococcifera
(first described by Desfontaines in Algeria) is the
most common Oak of Syria and Palestine; excessively variable, not only in its leaves, but in the
form and size of its acorns and cups, and even in
the raggedness of these last.   It is most commonly
(like Quercus coccifera) a hard scrubby bush, but
sometimes a low tree, and in *one* instance a very
large tree.   "Abraham's Oak" in the plain of
Mamre (traditionally said to be the very tree under
which Abraham entertained the Angels) a tree of
vast size and age, is, according to Hooker, Quercus
pseudococcifera.   He exhibited a sketch of it.   I
understood him to say that acorns from it had been
brought to Kew, and young plants raised from them.
Quercus Ægilops and Quercus infectoria are also
common in Syria and Palestine, and very variable,
even in their acorns as well as their leaves.   Quercus
Ægilops never a bush, but always a low or middle-
sized round headed tree, growing generally in a
rather scattered manner.   Mr. Ball, Bentham, and
some others, discussed the questions relating to
these Oaks, and afterwards Mr. Ball shewed us a
great number of specimens of Quercus coccifera and
pseudococcifera, from his own herbarium, illustrating the geographical range of each, and their

variability. In their most typical forms they appear distinct enough, especially in general appearance, but both are very variable, and it becomes very difficult to draw the line between some of their varieties.

Bentham told me he had ascertained that Dr. Gray was right, as to Du Chaillu's plates of the Gorilla and other apes being copied without acknowledgement from the illustrations of Isidore de St. Hilaire's paper in the Archives du Museum : and not only so but that he had copied the French figure of the Chimpanzee as a representation of his supposed new species. I have since myself looked at these plates, and come to the same conclusion.

June 22nd.

We spent the day with the William Napiers at Sandhurst, and in returning at night we saw the glare of a great fire lighting up the sky in the east. This was the great and terrible fire in Tooley Street on the south side of the river, near London Bridge, which proved so fearfully destructive, and in which poor Braidwood lost his life. When we got to Waterloo Bridge, we had a good view of it, and it was a grand and awful sight.

June 23rd.

Visited the Zoological Gardens with some of the Horner party. The gardens looking very pretty, and many fine and curious animals in them. The Japanese Pigs most grotesque looking creatures, with

1861. large broad pendulous ears, like those of a hound, short broad muzzle, reminding one of a Mastiff or Bull-dog, a face deeply furrowed or wrinkled in such a fantastic way that it appears tattooed. Young of this breed have been reared in the gardens, and they are thought likely to be a useful addition to our races of swine. The Wart Hog from Africa, a most hideous and monstrous looking creature. Young Peccaries, born in the gardens. The Babiroussa, a great rarity from Sumatra or Borneo, I forget which—of a remarkably light active form, and very pretty for a *pig;* seems very healthy, lively and active ; its tusks very curious. A young Giraffe born here on Whit Monday last ; a beautiful animal. Several thriving specimens of the hybrid race between the Hare and the Rabbit, which is exciting so much attention and curiosity ; they are however more curious in their history than in outward appearance. Young Emus, lately hatched in the gardens from eggs which were sent from a distance (from somewhere in the country) and sat upon by a male Emu ; the young birds curiously striped longitudinally with dark and pale brown. The "Mooruk," so called, one of the Struthionidae, so like the Cassowary that the difference is not evident. The Pinch Monkey, Hapale Œdipus, from the forests of New Granada, one of the smallest and most beautiful of the monkey tribe ; even smaller than the common Marmozet; never before, I believe, seen alive in Europe. The Sloth and the " Flying Foxes" (Roussette Bats) flourishing.

Minnie and Sarah dined with us, and we took them to the Evening Exhibition of the British Institution, for which I had tickets. It was a pretty sight. I had already seen the Exhibition by day-light and admired it; some of the pictures looked better by artificial light, others less well. The collection of Sir Joshua Reynolds' works is large, and very fine. Of the other pictures, those I particularly admired are : — Rubens' Second Wife ; a most beautiful portrait,—two very noble and ad-mirably well painted Venetian portraits, by Moroni ; "A garden scene with figures," by De Hooge ; a delightfully natural, home-like picture of the more refined aspect of Dutch life—the life of Dutch ladies and gentlemen ; with all the Dutch fulness of detail and unmistakable reality, without the usual coarse-ness. Also an exquisite little Cuyp, and some fine portraits by Vandyke. One or two others struck me rather by their singularity than their beauty. The Flight into Egypt, by Zurbaran has a great deal of effect, but the personages are all Spanish gipsies. The family of Sir Peter Lely painted by him ; a subject oddly treated, some of the ladies half naked. Henry, Prince of Orange, with his family and Secretary by Gonzalez Coques, a curious picture.

June 26th.

Went to the Geological Society and arranged with Rupert Jones about the illustrations for my paper on the Nagpore Fossil plants.

We went to Kew with Minnie and Sarah and
spent some hours in the gardens, which are in great
beauty. Several of the Palms and other tropical
trees in the great hot-house are really magnificent
and they are now shown to much greater advantage
than before, by being plunged in the ground instead
of standing in tubs. The Ferns are in a most
flourishing condition and wonderfully various and
beautiful : the gigantic Angiopteris and some of the
Marattias, especially magnificent ; some fine Fern
trees, Alsophilas and Dixonias but scarcely equal to
what I saw formerly at Berlin. Several species
of Glichenia (which I think one seldom sees in
cultivation) very flourishing ; the Glichenia flabellata particularly beautiful. Of smaller Ferns,—Da-
vallias, Adiantums, Aspleniums, Polypodiæ and
Acrosticheæ, in endless variety and beauty. The
magnificent Birds-nest Asplenium from Australia is
now separated as a species from the Indian one
under the name of Neottopteris Australasica.

June 28th.

Went to the British Museum and looked through
part of the collection of minerals.

In this month of June there were two remarkable
deaths ; the Chancellor (Lord Campbell) died suddenly in the night, at the age of eighty. A peculiarly
fortunate man, in my opinion, to close a useful and
honoured career at such a mature age by a speedy

death, escaping all the miseries of lingering decay 1861. and helpless suffering.

The other remarkable death was that of Mr. Braidwood, the chief of the Fire Brigade, who perished while gallantly doing his duty at the terrible fire near London Bridge on the 22nd: a death as noble as that of an officer at the head of his regiment in battle.

The new names for the three principal sections or divisions of the Church of England : — Attitudinarians ; Latitudinarians ; Platitudinarians.

An anecdote of Serjeant Merewether.—The late Chancellor, meeting him one day,—not having seen him for some time before, — said : "Why, Mere- "wether, how fat you are grown ! You're as fat as " a porpoise ;"— " Then, my Lord, I am the fitter " company for the *Great Seal*."

The new Lord Chancellor, Sir Richard Bethell, (now Lord Westbury), is noted for saying bitter and disagreeable things :—While the legal appointments and arrangements consequent on his elevation to the Chancellorship were under discussion, some one remarked that it would be unfair to make Sir Roundell Palmer Attorney-general over Sir William Atherton's head. "Over Sir William " Atherton's head ? You think, then, he has a " head ?" said the mild Chancellor.

We left London on the 4th July, taking with us my niece Fanny; went to Brighton and spent twelve days there pleasantly enough.

On the 16th of July we went from Brighton to Hamblecliffe, on the Southampton Water, and spent two days there with the Frederick Freemans and my dear old friend Lady Napier (the widow of Sir George). It is a pretty spot; the views of the estuary with the wooded and cultivated banks, and the hills of the Isle of Wight in the distance, very agreeable. The 19th of July we went from thence to Oaklands, where we remained till the 24th, spending our time very pleasantly with Lady Napier, dear Minnie, and charming Sarah.

On the 20th we went with Minnie and Sarah, and saw the beautiful gardens and grounds of Leigh Park, formerly Sir George Staunton's, but lately sold to a Mr. Stone.

The collection of tropical plants in the hot-houses is one of the finest that I have seen in this country (Kew excepted), both for the number of plants and for their beauty and vigour of growth. I noticed particularly a magnificent Allamanda, covering a great space, and loaded with gorgeous gold-coloured flowers like those of an Echites; Nelumbium speciosum, very fine, with many of its curious seed vessels; Victoria Regia in flower; Rice in flower; those grand Ferns, Angiopteris erecta, and Asplenium marginatum, in high perfection, and many other beautiful Ferns.

24th July. We went to London; had a good

talk with Edward, who is very lately returned from 1851.
Greece, highly delighted with his travels there.
The 25th we returned home. The 26th to Mildenhall. The 27th the presentation of musical instruments to the Mildenhall Volunteers, and speech-
ifying on the occasion by Fanny and myself. In the
evening of that same day we returned home again.
The 30th we started off again and went to Mr. Mills
at Stutton, seven miles from Ipswich. The 31st
I was engaged all day at the Assizes, where I
was foreman of the grand jury. Next two days
were spent very pleasantly at that charming place
Stutton. It is a gem of a place. I was pleased
to find that the beautiful evergreen and other trees
and shrubs had not suffered materially from the last
winter, except that the famous tall Cypress is much
cut on one side.

On the 2nd of August, Mr. Mills took us to that
beautiful place Woolverstone, where we had luncheon
with Mr. and Mrs. Berners, and saw the delightful
grounds—far surpassing, to my mind, everything
else of the kind that I have seen in Suffolk. Mr.
Berners showed me the Ferns, in which he seems to
take great delight, and of which he has a fine
collection, both of hardy and of hot-house kinds.
The Fernery on the 'Cliff' is flourishing, and
several exotic species have stood the cold of last
winter without apparent injury; Onoclea sensibilis,
Struthiopteris Pennsylvanica (Germanica), Lomeria
alpina, and what is very remarkable, even Woodwardia radicans. The new Fern hot-house contains
a good number of beautiful and interesting species

1861. and very fine specimens. I saw there also Bou-
gainvillea Brasiliensis in flower.

On the 3rd of August we returned home.

Finished the second volume of Buckle's "History
of Civilization." It is a book well worth reading.
Having in the first chapter disposed of the Civilization of Spain, he devotes the rest of the volume to
that of Scotland. I do not think the Scotch will be
well pleased with what he says of them. The
summary of the general History of Scotland, in his
second and third chapters, does not appear to me
particularly well done; except that he shews perhaps
more distinctly than others have, how cruelly that
country suffered, in its civilization as well as in
its material interests, from the barbarous incursions
and ravages of the English.—Much the most
curious part of the book is the chapter in which
Buckle gives a copious and detailed account of the
intellectual state of Scotland during the 17th century, and shows what was the nature, extent, and
effects of the influences exercised by the preachers
and theologians of those times. This is exceedingly
curious, and most especially so, the copious extracts
from the sermons and theological tracts of that age
by which he supports his views. It is altogether a
wonderful picture of gloomy superstition and of
ecclesiastical tyranny, searching and penetrating
into every relation of life.

We went, a large party, into Bury, to witness the opening of an Exhibition of Art and Antiquity, in the rooms of the Athenæum there. The object was to assist the funds of the Athenæum Museum. Lord Arthur Hervey, who had been indefatigable in forwarding and getting up this exhibition, opened it with a very neat, clear, sensible and agreeable address. The exhibition was very creditable : many pictures, chiefly portraits, and in particular, a very good show of portraits connected with Suffolk families, or illustrating the history of the county ; many interesting ones from Ickworth, Hardwick, and Hengrave. I sent four or five. There was also a pretty display of china and nick-nacks, besides some good collections of small objects of antiquity. Among the most curious articles was the arrow-head which was found imbedded in "St. Edmund's Oak," at Hoxne, some years ago, when the tree was shattered to pieces in a storm. The tree was certainly of very great age, and tradition always affirmed that it was the identical one to which St. Edmund the Martyr was bound when he was shot to death by the Danes.

LETTERS.

My Dear Katharine.

I thank you very much for your kind letter, which gave me great pleasure ; and I had intended to answer it sooner, but a house full of visitors and

1861. the constant fine weather, which tempts one to
be much in the open air, are great impediments
to letter writing, and indeed to all *serious* occupations. I get hardly any reading done in the day;
but I generally—not always—manage to secure one
hour in the twenty-four for my herbarium. We
have had glorious harvest weather; the finest that
could be desired, for the wind accompanying the
sunshine has contributed to the perfect ripening
of the ears, and the heat has not been so great as to
be oppressive to the labourers. I do not think I
ever experienced more enjoyable summer weather
than we have had, with very few interruptions, since
the beginning of this month. The wheat in this and
the adjoining parishes is now almost all gathered in.
The experienced all tell me that it is one of the
finest crops they ever remember, and has been got
in in the finest condition. It is a great blessing.
And now—as human nature, at least agricultural
human nature, is never long satisfied—we are beginning to want a little rain for our grass and
turnips.

We are, as you are aware, a large family party
here,— not less than twelve (including children), of
the Horner blood or connexion; and we have also in
the house at present, Mr. and Mrs. Bowyer, very
agreeable people (did you see them at Folkestone?)
—and Mr. Zincke, whom I think pleasant. It is a
great satisfaction to see Mr. and Mrs. Horner
looking so well, and so able to enjoy life; Mr.
Horner so indefatigably active in mind, and Mrs.
Horner so full of activity in every way. It is a great

pleasure too, to have our sisters with us. I have at last succeeded in getting pretty well acquainted with Annie and Lizzie ; they are very nice children when one once comes to know them. I was delighted to hear such a good account of your dear little people, and I beg you to give them my especial love, and a great many kisses for me to each of them, more especially to Rosamond and Arthur ; and tell them not to forget Barton or Uncle Bunbury.

Our garden and pleasure ground are now in great beauty, full of flowers, the conservatory very gay, and the Ferns in the greenhouse very fine and flourishing.

I have lately added to my collection of living Ferns, Onoclea sensibilis, Cystopteris bulbifera and two or three others, given me by Mr. Berners, when we visited him at the beginning of this month.

His park at Wolverstone, on the river Orwell, below Ipswich, is the most beautiful that I know in Suffolk, he is very fond of gardening, and he has formed a charming Fernery on a high and steep and thickly-wooded bank, almost a cliff, facing the north, above the river, where almost all the British Ferns, and not a few foreign ones flourish in the utmost beauty and luxuriance. It is a most choice situation for a Fernery.

You speak of the British Roses, which Bentham has reduced to five ; Babington made nineteen species even in his first edition, and I have no doubt he has since increased the number. The five are very clear and well-marked, but if one

1861. pushes the division still further, it is very difficult
to stop.

I hope you are all enjoying your stay at Sheilhill,
Ever your very affectionate Brother,
CHARLES J. F. BUNBURY.

JOURNAL.

September 12th.

We have had our house full ever since the 1st
August. Most part of the time a large family
party: amongst others Mr. and Mrs. Horner,
and the Pertzs, also our old Madeira friends Sir
Frederick and Lady Grey, very agreeable and
accomplished people. Since we saw them in
Madeira they have spent some years at the Cape,
and have also visited Mauritius and Bourbon and a
much less known island, that of Johanna in the
Mozambique Channel: and in all these places they
botanized.

Finished the first volume of the Diary and Correspondence of the first Lord Auckland. I had read to
Fanny, while at Brighton, a considerable part of
the second volume, comprising his Diary (written
for his Mother) while he was Minister at the Court
of Madrid, which is very entertaining. The first
volume contains—1st, a short notice of Lord Auckland's life:—2nd, his correspondence chiefly with
Lord Loughborough, about the Coalition which
Lord Loughborough had a principal share of bring-
ing about :—3rd, his offical correspondence while he

was employed at Paris, as Envoy extraordinary, 1861. in negociating Mr. Pitt's Commercial Treaty, and afterwards in those discussions of France to which the affairs of Holland gave rise. This part of the volume is politically important, but heavy and tedious:—4th, letters written to Mr. Eden by his friends from England, while he was negociating in France. In these there is much that is entertaining, especially the letters from Mr. Storer. In the earlier part of the book, there is a curious account of the punishment of Madame la Motte for the "Diamond necklace" affair. — There are many letters from Gibbon's friend Lord Sheffield, (Holroyd), and they give me a very unfavourable impression of his mind. He seems to have been a fussy, self-important, bustling politician of the most common-place kind, believing himself to have been a great authority on matters of finance and commerce, and constantly advocating the most narrow, illiberal, and ingenerous views on those subjects. Lord Sheffield evidently thought himself a very important personage, and little imagined that after two generations, he would be known only as the friend and editor of Gibbon. It appears that France was very near going to war with England, in 1787, upon the Dutch dispute. Eden, who was on terms of personal friendship with several of the French Ministers, had great difficulty in preventing an open rupture ; and it appears that, though the Archbishop of Toulouse, the Prime Minister, was for peace, and succeeded in maintaining it, there was a strong party in the French Ministry that was

1861. eager for war; indeed the conclusion of the business was so like a submission on the part of the French, it was no wonder they were indignant. The dreadful state of the French finances was no doubt the great peace-maker; but some, it appears afterwards, thought that a Foreign war might have averted the crisis of the Revolution. It could, however, have been only a temporary respite.

I read last winter, in "Lord Malmesbury's Memoir," very copious details about this struggle between the French and the Anglo-Prussian parties in Holland; it is curious to read another account of the same transactions from so different a point of view as Lord Auckland's. Neither of the two English negociators appears to have thought at all favourably of the others' exertions.

Barton.

September 18th, 1861.

My dear Mary,

I have no doubt that Charles Lyell is, as usual, very busy, and therefore I am unwilling to trouble him with a letter. But I shall really be very much obliged to you if you can ascertain and write me word, what is his opinion of the Middle-class Examinations of the Society of Arts It is proposed to establish at Bury a Local Educational Board in connection with the Athenæum there, in order to enable those who attend the classes to take part in those examinations; I have been asked to be a member of the Board. I do not know enough of the matter to judge whether the examinations are really useful, and as I have no doubt that Lyell has paid some attention to the subject, I should like extremely to know his opinion. If you can spare time to write me a line about this, I shall be very thankful to you.

I am heartily glad to see such a glorious sunshine on the day when our dear friends* are to cross the water. It is indeed a delicious morning ; one of those peculiarly fine, still and clear mornings of autumn which are as delightful as any in the year ; and if it is as still at Folkestone as it is here, the travellers will have a capital passage. I hope and trust their expedition will be altogether prosperous and enjoyable.

Edward is here, in unusually good health and spirits.

With much love to Lyell,

I am ever your very affectionate Brother,

C. J. F. BUNBURY.

JOURNAL.

September 22nd.

Finished the second volume of Lord Auckland. The first two hundred pages (two-fifths) of it, indeed, I had read to Fanny while we were at Brighton. The remainder of the volume consists of unofficial correspondence containing much that is curious and entertaining. It relates first to the time of the

---

*Mr. and Mrs. Horner and their daughters Susan and Joanna on their way to Italy.

1861. great party struggle about the Regency during the illness of George III. in 1788, and the beginning of '89 ; the correspondence gives us a lively impression of the hopes and fears, anxieties and expectations of politicians during that critical time.   Secondly, the French Revolution from its first dawn to the beginning of the war in 1793.   In the earlier part of this time the numerous letters of M. Huber have some interest because he was not only a resident in Paris, but was an intimate friend or connection of Necker, and seems to represent more particularly *his* views and feelings.   The latter part relating to the beginning of the Reign of Terror is curious chiefly as showing how little expectation of war there was on the part even of those more closely connected with the English Government till almost the very eve of the actual struggle.

Read the first chapter of Gibbon's " Decline and Fall."

September 23rd.

I found a specimen (the first I have ever seen here) of that rare bird, the Hawfinch or Grosbeak, Loxia coccoathraustes, lying dead in the arboretum ; perhaps killed by the violence of the wind.   It is very well described and figured in "Yarrell's British Birds."

Read the second chapter of " Gibbon."

September 26th.

Read the third chapter of "Gibbon's Decline and

Fall," These three famous chapters, the first three 1861.
of his great work are in truth masterly ; they are a
magnificent picture of the Roman Empire in its
prosperous and brilliant days.

LETTERS.

Barton,
October 1st, 1861.

My dear Lyell,

I have received from Boone, Prof. Heer's
fine work, also the "Recherchés, etc." by Heer
and Gaudin, and a book lent me by Susan. The
Flora Tertiaria Helvetiæ is indeed a grand work,
and I am very glad to have it, and promise myself
much interesting occupation in working gradually
through it, judging as well as I can of the validity of
Heer's determinations and making out the principles on which he has proceeded. I am very much
obliged to the Professor for his very liberal present
of the "Recherchés" ; pray when you see him, be so
good as to give him my hearty thanks for it. I have
begun to read it with much interest ; it seems to be
full of curious matter. The variety of plants that
have been found fossil in the Swiss tertiaries is very
extraordinary, though I dare say the number of
species would be considerably reduced if they were
dealt with after the manner of the Hookers and of
Bentham. Still this would hardly affect the variety
of *forms* or the number of families that seem to be
represented. It can hardly be doubted, and I think
that the flora of that country in the Miocene time
was much richer than that of any part of Europe at

1861. present, at least so far as trees and shrubs are
concerned, and perhaps even richer than that of
any of the temperate districts of America.   It
appears to me however that there is an element of
uncertainty in all those comparative statistical
details into which Heer enters so copiously in the
early part of this essay ; and that is the impossibility
of knowing how many species have perished without
leaving any trace.

Heer admits that nearly all the herbaceous plants
have probably thus disappeared ; and how are we to
estimate their proportional number ?   They *may*
have formed as small a proportion as in New
Zealand, when nearly one-third of all the flowering
plants have woody stems ; or they may have come
nearer to the European proportion.

I read with interest your letter of the 22nd.   As
touching the Planes—I am not myself acquainted
with the Platanus acerifolia, but I find that Loudon
makes it a variety of *Orientalis*, and it is said to be
a native of the " Levant."

The leaves of the fossil Platanus *aceroides*,
judging by Heer's figures, appear much more like
those of *occidentalis* (the American species) than of
orientalis.

There is no doubt that the Miocene tertiary
fossil flora of middle Europe has many striking
analogies to the existing flora of North America,
as shown particularly by its Taxodium (very close
indeed to the Deciduous Cypress), its Liquid-
ambar, its Tulip-tree, its abundance of Maples and
Walnuts, and some of its Oaks.   But on the other

hand, the Glyptostrobus Europæus which appears 1861.
to be one of the most common and general
plants of that age, and one of the best made out,
has its living representative in a Chinese and
Japanese tree ; the Cinnamomums are truly
Eastern forms ; the Proteaceæ are at present
equally strangers to Europe and to North
America ; and there are very many other
anomalies. Indeed what has all along puzzled
and staggered me with respect to these tertiary
plants, as determined by Heer and his school
is that the vegetation of Europe in those
ages would seem to have been made up of a medley
of forms, setting at defiance all known facts in
botanical geography ; forms now characteristic of
the most different and distant countries. I made
some observations on this matter in reviewing
Wessell and Weber on the brown coal in number 57
of the Journal.* There is a child's book called "Le
Robinson Suisse," (a great favourite of mine in
former days) in which the author has brought
together into one island for the comfort and
amusement of his heroes, the most remarkable
animals from all parts of the earth. The Miocene
flora of middle Europe would appear to have
been somewhat in the same miscellaneous style. I
do not at all mean to dispute Heer's determinations ;
I have not gone deeply enough into the question
but " 'tis strange—'tis passing strange." At the
same time I must admit that this mixture of forms

* Geological Journal.

1861. is only contrary to *analogy;* that we have hardly yet discovered anything that can be called a positive law in botanical geography; we do not know why certain families of plants are apparently confined to certain regions.

The question wether some or many of these supposed extinct tertiary plants may not be specifically identical with recent ones, is very curious and unfortunately almost impossible to solve satisfactorily, while naturalists are so little agreed as to what *is* a species. Heer's Swiss Woodwardia (a magnificent plant) is certainly excessively like our old friend the radicans; still I think its distinguishing character is such as *(if constant)* would be considered by the generality of botanists as sufficient for a specific distinction in the case of a recent plant; therefore the Miocene Woodwardia may perhaps be allowed to stand as a species till intermediate forms shall be discovered. I think I mentioned to you that the Woodwardia radicans, though found in many distant countries, seems to be less variable than the generality of Ferns.

I am very sorry to find that you and Mary cannot come to us while the leaves are on the trees, and Barton in its best looks. It is very unlucky that you can never come but in winter. I assure you it is a mistake: the country, and I think this place especially, really looks much prettier in summer or early autumn than in winter.

We have the finest weather imaginable, and have had now a long enjoyment of it. It is altogether a wonderful season. One is so constantly tempted out by the beauty of the weather, that it is difficult to settle down to indoor pursuit.

I was extremely glad to hear of the safe arrival of our travellers at Turin, and that they were so well and enjoyed their travels so much. Fanny read to me this morning a most agreeable letter from Susan, sent to us I think by Mary.

I am reading "Motley's United Netherlands," and am just now deep in the siege of Antwerp which is as interesting as any romance. He has an extraordinary power of narration.

With much love to dear Mary, I am ever,

Your very affectionate,

CHARLES J. F. BUNBURY.

JOURNAL.

October 3rd.

The autumnal colouring of the foliage here is more beautiful this year, I think, than I have ever seen them before, owing I suppose to the very dry and sunny weather we have enjoyed almost without interruption for the last two months; the leaves have changed colour very early, and their tints are peculiarly rich and vivid. A scarlet Oak near the library windows is most gorgeous at present with the various glowing shades of red of its leaves; seen in a mass, at a little distance, the whole appears of a deep red, but approaching nearer, we see it most beautifully variegated with shades of crimson, scarlet, orange and green. A Liquid-ambar at the

1861. west side of the lawn, has assumed a lighter and still more vivid crimson, almost rose coloured. Two American Oaks (either rubra or one of the nearly allied species) on the lawn, had turned, not red, but a rich golden brown. The tall Sycamores opposite the west front of the house, look far more brilliant than I ever saw them before, variegated with broad patches of a bright yellow, which looks quite golden in the sunshine. The Beeches are finely coloured, and even the Oaks are changing: the various rich tints of the Rhus typhina, Rhus cotinus and Euonymus and many other shrubs contribute to the effect; and the way in which all the various shades of colour of the foliage are blended and harmonized, is exquisitely beautiful. It is remarkable (and the same thing has been remarked by Mrs. Rickards) that, during this very fine summer and autumn, butterflies have been few. I have seen hardly any except the common white, and the beautiful Atalanta or scarlet Admirable. This latter kind was numerous for a few days in the early part of September, and I have seen it occasionally since. Wasps have been very numerous and troublesome.

October 8th.

We went to the Bury Athenæum to hear Lord Arthur Hervey's Lecture, the opening one of the season, on the "Dissolution of the Monasteries in England;"—excellent.

LETTER.

Barton,
October 21st, 1861.

My Dear Katharine,

The Freemans are with us still, but Lady 1861.
Louisa Kerr, who is an accomplished and agreeable
woman, went away a week ago. Augusta Freeman
is charming,—one of my great favourites. It has
been delightful to hear of our friends getting on so
successfully with their travels, and meeting with so
much to interest and gratify them, and especially of
Mr. and Mrs. Horner being so well and strong.

With so much company as we have had, I have
had not much time for reading. I have read over
again the 4th volume of "Cosmos." Now good-bye
dear Katharine, with much love to your husband
and children,

Ever your very affectionate Brother,
CHARLES J. F. BUNBURY.

JOURNAL.

October 27th.

Finished the first volume of "Motley's History of
the Netherlands." This thick volume of five hundred
and thirty-two pages, contains the history of two
years,—from the murder of William of Orange in
July 1584, to the end of the underhand negociations
for peace with Spain, in the summer of 1586 : and
notwithstanding the unquestionably great ability
with which it is written, and the interesting and

1861. critical time to which it relates, I confess that on
the whole, it is, to me, rather tedious. The
memorable siege of Antwerp by the Prince of
Parma, and its gallant and all but successful defence
by St. Aldegond (June 1584 to July 1585) is indeed
admirably well told, and is as interesting as any
romance. Mr. Motley's minuteness of detail is here
not at all excessive or cumbrous; every detail
heightens the general effect. He is great also in
picturesque and effective descriptions of prominent
characters: that of Philip II. influencing the fate
of all Europe from his writing table in the gloomy
recesses of the Escurial, is masterly; those of the
"three Henries" of France — Henry of Valois,
Henry of Guise and Henry of Navarre, likewise
admirable. St. Aldegond, Sir Philip Sidney and
the Earl of Leicester are also sketched with great
effect. But the negociations of the States, first
with France, and afterwards with England, the
whims and fluctuations of Queen Elizabeth and
the dark and dirty intrigues of some of her
Ministers with the Agents of Spain, are related with
a minuteness which makes these parts of the
volume, to me, not a little tedious.

————————

October 30th.

We went to the Bury Athenæum to hear Dickens,
who read selections out of "David Copperfield:"
I was exceedingly pleased. It so happened that I
had never heard him before. — and certainly I
never heard reading equal to it—Such clear and
agreeable tones of voice; such force and variety of

expression, without anything forced, overstrained, 1861.
or inclining to extravagance. In the pathetic and
impressive scenes, the story of "Little Emily" and
the death of Steerforth, he was admirable, but in
the comic portions he was superlative. The reading
lasted two hours, and I was sorry when it was
over.

October 31st.

Read in the second volume of "Motleys United
Netherlands," the story of the fight near Zutphen,
in which Sir Philip Sidney received his mortal
wound. It is one of the most interesting narratives
I ever read. It is as vivid, animated and picturesque
as any of Walter Scott's battle pieces, in short,
quite masterly; and what a fight it was! Though on
a small scale, and not attended by any great results,
it was I think as heroic as Agincourt or Albuera.
About 550 Englishmen making head for hours
against upwards of 3,000 Spaniards of the best and
choicest troops of Parma's veteran army, repeatedly
driving them back, and retreating at last only
because fresh reinforcements came up to the
Spaniards, while they through Leicester's unskilful
arrangements were left unsupported. The English
were led by some twenty gentlemen volunteers of
some of the noblest families in England, by whom in
fact the fight was chiefly maintained; "Essex and
"Audley, Pelham, Stanley, Russell, both the
"Sidneys, also the Norrises, Lord North and
"Lord Willoughby."

LETTERS.

Barton,
November 4th, 1861.

1861.   My Dear Mr. Horner,

It has been a great comfort to me, as well as to Fanny, to hear such excellent accounts of you and Mrs. Horner and our dear sisters, your travelling companions; that you made out your journey so satisfactorily, that you enjoyed so much what you saw and above all, that your health was so good.   I trust, and most heartily wish, that the same good health and good fortune may continue to attend you the whole time you are abroad, and that we may see you next summer all the better for your foreign travel.

I have read with great interest and pleasure all the portions of your journal as they have been successively sent to us; and especially the journey from Genoa to Pisa and Florence, and with your geological observation.   I know that country well, and hope I may see it again before I die.

I imagine Collegno's geological map is rather antiquated; I think he marks the schists of the Genoese mountains as cretaceous, whereas Murchison compares them to the *Flysch* of Switzerland, which is Eocene.   In fact those Genoese rocks have been travelling *upwards* through so many stages that I hardly know where they will stop!

In some of the defiles on the North side of the Genoese mountains, on the newest post road from Turin to Genoa, those schists as I remember, look

very like some of the old Devonian slates, and their 1861.
laminæ are most wonderfully crumpled. I daresay
you observed the great tract of serpentine (containing abundance of diallage in large flakes) in that
mass of mountains between Sestri and La Spezzia,
and the wild colouring of lurid green and dusky red,
which it gives to many of the bare hill sides.

I was struck by what you mentioned about the
metamorphic rocks, and the granite bursting through
miocene strata near Campiglia. I have no doubt
you will find much that is interesting in the
museums of Florence, and hear much from the
learned men to whom you have been introduced
under such good auspices. Professor Parlatore was
not at Florence when we were there in '48, but I
made his acquaintance a few years ago at the
Linnean Society in London; pray be so kind as to
remember me to him when you next see him.
Targioni Tozzetti, who showed me great attention
and kindness in '48, is, I am afraid since dead; he
not only showed me the Botanic Garden, but
procured for me access to the great herbarium;
it was even then a very fine collection, (and as far
as I saw) in remarkably good order, but not com-
parable in extent to those of Kew or Paris but that
was before the acquisition of Mr. Webb's herbarium;
I shall be curious to know what you think of the
mineral department of the Florence Museum.

Florence is altogether to my mind the most
beautiful and the most agreeable of all cities—at
least of all that I have ever seen; the most interesting too, with the exception of Rome. I almost envy

4 X

1861. you for spending the winter there.   And how
delightful it must be to see that glorious country in
a state of orderly freedom.

For ourselves we have been perfectly stationary
ever since you left us, but have hardly been alone
any part of the time.   We had intended to be at
Mildenhall this week, but poor Frederick Freeman
has fallen very ill while staying with us, and so he
and his wife are detained here, and consequently we
likewise.   Fanny is, I think, quite as well as
usual.

We have had a most beautiful autumn ; the finest
I ever remember: almost uninterrupted fine weather
throughout September and October.   Then we
were carried at once, with one leap into the middle
of winter.   When I got up on Saturday morning,
the 2nd, it was *snowing* furiously, with a violent gale
from the N. W., and though the snow did not lie
long, it was so heavy in the morning as to break
many boughs.   The wind also did much damage to
the trees in our park.   On the Northumberland and
Yorkshire coasts, the gale of Saturday morning
appears to have been tremendous and very destructive, especially at Scarborough.

Pray give my hearty love to Mrs. Horner, Susan
and Joanna, and believe me ever,

Your very affectionate Son-in-law,

CHARLES J. F. BUNBURY.

Wednesday, November 6th.

JOURNAL.

A very fine day, with a slight frost in the morning. Notwithstanding the snow which fell on the 2nd, there has not yet been frost enough to kill the leaves of the sweet-scented Verbena (Lippia) in front of the house, and the Amaryllis Belladonna is still in blossom, in the border before the conservatory. The gale of Saturday has done much damage in the Park here, but none of any consequence in the arboretum. On the east coast the storm appears to have been tremendous, and sadly destructive to shipping and to human lives.

We dined with the Arthur Herveys; met the Dean of Westminster (Trench) and his daughter.

I read a little book which Dean Trench had given me the evening before, entitled "A Visit to Germany," 1799-1800, — being a portion of his Mother's journal kept at that time, which he has had printed but not published. It is very entertaining, written with great vivacity and acuteness, and showing considerable powers of observation. The lady (she was a young widow at the time) seems to have had excellent introductions, and her sketches and anecdotes of several noted and historical personages

4 X 2

1861. are very lively;—those of Nelson and Lady Hamilton, particularly striking though very unfavourable.

In the evening went to the Bury Athenæum, and heard the Dean lecture on the poetry of Wordsworth. The matter and language of the lecture were excellent; the manner and tone not so agreeable.

November 13th.

Lord Arthur Hervey and his daughter Sarah (a charming girl) with the Dean of Westminster and Miss Trench, came to luncheon with us and to see the pictures and library; the day was excessively wet, they could see nothing outside the house.

Our impression of the Dean is very agreeable.

November 18th.

Finished Lord Stanhope's "Life of Pitt," at least the two volumes hitherto published.

I have been a long time about it, though it is very easy reading. Fanny at first began reading it aloud to me, while I was sitting to Mr. Boxall, before the end of August; and for a long time we tried to go on with it in that way, sometimes the one, and sometimes the other of us reading it aloud. But as we have had company in the house almost constantly since that time, our progress was very often interrupted, and was very slow; till at last we gave up that plan, and were content to read it separately to ourselves.

This book is, as I said, very easy reading, and 1861.
very pleasant reading. It has much the same
general character as his " History of England," not
brilliant, not profound or philosophical; but
eminently the work of a gentleman—of a man of a
mild, candid, eloquent, accomplished mind. The
style clear and easy, not forcible nor picturesque. I
have been disappointed however in one material
point ; I have learned from the book very little that
I did not know before. Considering that Lord Stanhope through the connexion of his family with that
of Pitt, must have had the use of whatever materials
there were for a biography, and that one cannot
suppose him to have been negligent in using them,
it is surprising how little of Pitt's *personal* history
this book contains.

It gives us, indeed, many of Pitt's letters to his
mother, which have a certain degree of interest, as
showing the warmth and steadiness of his domestic
affections ; but nothing more. Otherwise this, like
all the previous biographies of Pitt, is a history of his
public life ; a history of Pitt the statesman, rather
than of Pitt the man. I suppose this is not the
fault of Lord Stanhope, but belongs to the character
of his hero. I suppose that Pitt had as little of personal individual existence, separate from his political
life, as it is possible for any man to have.

There is scarcely any historical character about
which my opinion has changed so much as about Pitt.
Accustomed in my youth, of course, to look upon him
as the great champion and representative of Toryism,
in opposition to Fox, who was to be considered as

1861. the representative of Whiggism, I have long been coming more and more decidedly to the conclusion that Pitt was the wiser and more enlightened statesman of the two.

Lord John Russell's "Life of Fox" had settled this opinion in my mind almost as strongly as the present work of Lord Stanhope. Pitt was indeed, in some respects, in advance of his time ; especially in his commercial policy, which was truly enlightened, and worthy of a professed admirer of Adam Smith. His policy, whether domestic or foreign, after 1792, is certainly more questionable, but I think there is much to be said on both sides, and that it is by no means proved that the course recommended by Fox would have been safe, or even compatible with national independence.

November 20th.

We went to London.

November 21st.

We dined at the Lyells ; I went with Charles Lyell first to the Linnean Society, but we were not able to stay for more than the beginning of Darwin's very remarkable paper on some sexual peculiarities in the flowers of the Primrose ; as there were some papers to be read at the Royal Society which we were still more anxious to hear, and which had unfortunately been fixed for the same evening.

Walked home from the Athenæum with Louis Mallet, and had much talk with him on politics. He is a keen politician.

He talked of the War in America ; thought the Southern States would soon be obliged to yield, through their inferiority in pecuniary and material resources.

Concerning France, Louis Mallet had seen a letter received by Mr. Cobden from Michel Chevallier, the eminent political economist, in which he (Chevallier) expressed himself as satisfied that the Emperor is quite in earnest in his intentions of retrenchment, peace, and liberal policy. This however, is only an individual opinion. *Valeat quantum.*

LETTERS.

Barton,
December 1st, 1861.

My Dear Joanna.

The accounts we continue to receive of all of you are very delightful, but I have heard less of you individually than of any others of the party. Mr. Smith,* whom I saw in London, spoke with most warm gratitude of what you had done for Mrs. Clough. It was indeed like yourself. Poor Mrs. Clough ; I am very sorry for her. As for Mr. Clough,† I knew him so very slightly that it would be affectation to say that I regret him ; but he seems

---

* Mr. Samuel Smith, father of Mrs. Clough.
† Arthur Hugh Clough died in Florence, November 13th, 1861.

1861. to have been a valuable man.   There has been an interesting and well written notice of him in the *Spectator*, written, as I understand, by Mr. Hughes, the author of " Tom Brown."

Fanny and I are just returned from London where we spent ten days.   It was a great pleasure to see something of the Lyells, both of No. 15 and No. 53, but otherwise I cannot say that London appeared to me very enjoyable at this season of mud and fog and gloom.   However, I did great things in the way of paying bills, and Fanny rushed about with un-ceasing energy to all parts of the town, got through a tremendous quantity of business and of course knocked herself up.   On Thursday, the 21st, I attended   the   Royal   Society   Meeting   where we had an interesting geological evening.   Mr. Pengelly's account of the lignites of Bovey Tracey (in Devonshire) and Professor Heer's report on the fossil plants collected there, were read—both in abstract.   Heer's was particularly curious and im-portant.   It seems to me now clearly made out that those Bovey lignites are of true Miocene age, cor-responding to the greater part of the Swiss Molasse, and of the Brown coal of Germany ; and they appear to be the only example of that formation in the British Islands except that little remnant discovered by the Duke of Argyle on the top of the basalt in Mull ; unless indeed the Hempstead beds in the Isle of Wight should be considered of the same age. The prevailing plant in the Bovey beds is a coni-ferous, closely allied to the existing Wellingtonia (the mammoth tree of California) ; of this there are ample remains, wood branches, leaves, cones, and seeds, so that its nature is well made out; it is a new species, different from any known on the Continent, and has been called Sequoia Couttsia, the researches having been carried on (as you are perhaps aware) at Miss Coutts's* expense. Altogether, Heer makes out 49 species of plants of which about 20 are identical with known Miocene species of Germany and Switzerland, and the rest though considered distinct are nearly allied to Miocene species. But in some cases the materials for determination seem but scanty. The only trace of an animal that has been found in the whole formation is a single fragment of a beetle.

I am afraid Charles Lyell is working rather too hard; he does not allow himself sufficient relaxation. We are looking forward however to the pleasure of seeing him and Mary as well as Katharine and Harry and their children here at Christmas, and a great pleasure it will be. I hope you will all enjoy your Christmas at Florence. I daresay you find much to interest you in the Natural History Museum. I do not well remember what sort of a collection of fossils they have, but I daresay they are rich in the Tertiary shells of Tuscany; and the Minerals, I well remember are a fine collection.

The galleries of the Uffizi, and the Pitti Palace afford a perpetual feast. I beg you to give my most *tender* remembrances to Titian's Flora.

Do you know that some of the Cypresses now

---

*Baroness Burdett Coutts.

1861. growing at Barton were raised from seeds which my Father gathered off trees in the gardens of the Villa Capponi, when poor Henry Napier was living there. They are still alive, though much hurt by last winter.

As to politics, the only thing thought of now, in this country, of course, is the insult offered by an American frigate to our flag, and the question whether we shall have war; that is, whether the American Government will uphold what has been done by their officers. I have still some hope that the whole business may end in the overthrow of Mr. Secretary Seward, who seems to have been all along determined on a war with England; but the hope has become very faint since we have learnt the tone taken by the New York newspapers. The curious thing is that in making this seizure, the Northerners have virtually acknowledged the independence of the South; for if these Southern men were *rebels*, the others would have no more right to seize them on board our ships than the Austrians would have to seize Hungarians or Venetians on board American vessels.

(December 4th.) Beautiful bright weather for some days past, with moderate frost. Poor Frederick Freeman,* who has had such a tedious and painful illness, is much better, and seems at last to be really recovering.

Much love to Mr. and Mrs. Horner and dear Susan.

Ever your very affectionate Brother,

CHARLES J. F. BUNBURY.

---

* Frederick Williams Freeman, the stepson of Sir George Napier, had married the daughter of Capt. Henry Napier, and was staying at Barton this autumn.

JOURNAL.

Finished the second volume of " Motley's History of the United Netherlands." It has the same merits, the same faults as the first. I have already noticed his admirable narrative of the combat of Zutphen ; and the story of the Spanish Armada also is told with remarkable spirit and clearness ; but in general I found the book tedious—overloaded with excessive minuteness of detail. The author seems to me not to observe sufficiently the distinction between the materials of history and history itself.

The character of Queen Elizabeth certainly appears in a sadly unfavourable light in these pages; and as all that Motley says of her is founded on authentic correspondence at the time, it is difficult to see how her fame is to be vindicated. Her excessive and almost ridiculous avarice, her jealousy, her vanity, her caprice, her blind partiality to the Earl of Leicester, were most seriously hurtful to the interests of England and of Holland, and constantly exposed both countries to extreme danger ; nor does Burleigh appear to much more advantage. Though his faults were of a different nature from those of his mistress, he was equally a dupe to the craft and dissimulation of Farnese and of Philip. It is really frightful to think how narrow an escape England (and with her all Europe) had of falling a prey to Spain. They were saved, not by the wisdom of Elizabeth or her ministers, but, under God, by

1861. the gentlemen and seamen of England, and the sturdy mariners of Holland.

LETTERS.

Barton,
December 3rd, 1861.

My Dear Emily,

We returned a few days ago from London, where we had spent ten days. We did not see many people besides the Charles and Henry Lyells, but we went down one day to Twickenham, and saw the dear Richard Napiers, and found them I am happy to say, both well (all things considered) and in good spirits; and as they always are, very agreeable and full of kindness and warmth of feeling towards us. Emily* was with them, looking well and cheerful, though her hair is no longer grey but pure white,

Poor Frederick Freeman, who has had such a long and distressing and lowering illness seems at last to be really rallying. They talk of leaving us next Monday or Tuesday.

I have no doubt you are as much interested as we all are in the great and serious question which has in fact for the time superseded all other political questions : whether we shall have war with the Northern States of America ; that is whether their Government will uphold the act of Captain Wilkes in taking prisoners from under our flag. There seems now to be some reason to suppose that the American

* Daughter of Sir William Napier.

Captain did not act under direct orders from his 1861.
Government, but that he took on himself the
responsibility of an act which he felt confident
would be agreeable to his masters. I have little
doubt that Mr. Seward, the American Secretary of
State, is determined on a war with England; all his
speeches and acts have pointed that way; and it is
said he hopes to gain Canada as a *set-off* against the
Southern States. It is likely that some of the
members of the Government may be differently
disposed; but it is a difficult and anxious question
whether they will venture to disavow an act which
has already been so highly extolled by the American
newspapers, and which is sure to be agreeable to the
mob. Our Government appears to me to have
hitherto done very well in this matter; to have
shown spirit, promptitude and coolness; and I hope
they will be firm throughout and hold the honour of
England as high as in the old time. It will certainly
be rather deplorable if we are forced into a close
alliance with the slave States, but the responsibility
must rest on those who drive us into it. If we must
have war, I have no doubt we shall soon see plenty
of the good old high spirit springing up in this
country; and the feeling in Canada is said to be
excellent. It is rather striking to see how decidedly
the French newspapers appear to take *our* side in
this dispute. Edward says he has heard on good
authority that the French Government has for some
time been strongly urging ours to acknowledge the
Southern confederacy, and to break the blockade.
It is said that the stoppage of cotton is even more

1861.   seriously felt in France than in England, for though
the actual quantity used is much less, it is equally
essential to the prosperity of their manufactures, and
they have no stock in hand.   I am told that a large
quantity of cotton is used even in the Lyons
manufactures ; it is mixed somehow with the silk.
Louis Mallet, who is very well acquainted with our
manufacturing districts, says it is quite true that the
masters are rather glad than otherwise of a
temporary stoppage of the supply of cotton ; that
they had been over producing—glutting the markets
to such a degree, that they must soon have reduced
their rate of working, even if this war in America
had not happened, and this relieves them from the
responsibility of doing so.   They are said to have a
stock of manufactured goods on hand equal to two
years consumption of the whole cotton-wearing
world.

Before this disagreeable affair of the Trent
happened, the Lyells had been much pleased at the
news of the Northerners having taken Beaufort in
South Carolina, which they consider an important
advantage.

Charles and Mary Lyell are still sanguine as to
the final and complete success of the North, but I
think there are very few besides who take the
same view.   Louis Mallet told me that Sir James
Ferguson, who was an officer in our army, and
served in the Crimea, and who has lately been in
America and visited both armies, says that the
Southern officers are more experienced and more
soldier-like, and their men (on the whole) finer

men and better drilled and disciplined than those 1861. of the North ; but the Southern army is very inferior in equipments, military stores, and in short in *material*.

Fanny is pretty well, and I am as well as I ever am in winter. I need not tell you of the *Cottagers*, as I know Cissy writes regularly to you ; they all seem most flourishing except poor Henry, who is still frequently a prey to neuralgia.

Your very affectionate Step-son,

CHARLES J. F. BUNBURY.

JOURNAL.

December 4th.

Began to read the "Œuvres et Correspondance Inédites d'Alexis de Tocqueville," by Gustave de Beaumont.

December 7th.

Edward writes to me that the general impression in London seems to be, that war with America is inevitable, "though it may not be the *immediate* result of the dispatches forwarded to Washington. I am told that this is the tone taken by the American Minister here, Mr. Adams, who himself is a moderate and reasonable man, and that Russell, the *Times* correspondent, in his private letters by the last Mail, expressed the same opinion. No one believes that the Americans will give up the men (Slidel and Mason) or indeed that they *could* do it

1861. even if the Government and Senate were disposed
to do so, for the mob would be too strong for them ;
and it is certain that we shall accept nothing else."

George Napier is placed on the Staff in Canada,
with the local rank of Major-General, and is in fact
second in command in that Province ; so that he is
likely enough to see some sharp service.

On the 12th of December we went to Mildenhall
to stay.

_____

December 15th.

Read the first chapter of Mill on "Liberty."

_____

December 16th.

This day we received the sad news of the death of
Prince Albert, the Prince Consort :—unexpected as
well as most sad.   He died on Saturday night,
the 14th, but the news did not reach us (at
Mildenhall) till this Monday morning.   One's first
feelings, naturally, were those of sympathy for the
poor Queen, who has suffered such a terrible blow,
such an irreparable loss ; but the loss is also a public
and national one, a very wise, able, and good
man has been taken from us ; one who occupied a
post much more important in reality than in
appearance, and whose influence for good has
probably been more felt than understood.   The
feeling of sorrow for his death seems to be very
general and very sincere : a much deeper and more
earnest sorrow than has been felt for any other
Royal personage since the Princess Charlotte ; yet

it is probable that what the nation has lost will be 1861 more felt and more justly estimated hereafter than at present.

———————

December 20th.

Charles and Mary Lyell arrived. I learn from them that the death of the Prince was not so unexpected to those about him, as to the country at large; that the physicians in attendance on him had augured ill of the case, and been very apprehensive all the week, and indeed almost from the beginning of the illness. He appeared, they say, to have no vital power to struggle against the attack. The poor Queen would not believe in his danger, and even caused the bulletins for the public to be modified, so as not to convey a correct impression of his state.

———————

December 23rd.

On our way back to Barton from Mildenhall, Charles Lyell and I stopped to examine the very extensive Gravel pit on the brow of the uplands, between Icklingham and Lackford Bridge. This is the spot called "Rampart Hill," by Prestwich (in the Geological Society's Journal) where two of the much-contested flint hatchets were obtained by Mr. Warren. Great numbers of Roman coins and other works of art have also been found in and near this same locality. There was evidently a considerable Roman Station.

———————

4 Y

December 24th.

Dear Katharine and her husband and children arrived.

—————

December 26th.

I went with Charles Lyell to Ixworth, where we saw the remarkable collection of antiquities formed by Mr. Warren, a watchmaker, and a very intelligent man. We saw the two noted flint hatchets, supposed to be of the drift period, which had been found in the Icklingham gravel pits, and various other stone implements of a later time, and of more finer and careful workmanship, formed in various parts of the country, as well as abundance of articles of bronze. The quantity of Roman coins and other objects of art, which Mr. Warren had obtained from the immediate neighbourhood of Ixworth and Pakenham, is quite remarkable.

—————

December 28th.

Charles and Mary left us. Partial as they are to the Americans, they are naturally in great anxiety as to the prospect of a war between that nation and ours. They would consider such a war as an unmitigated evil, yet, Charles Lyell does not maintain that our Government is to blame in this matter, or that it ought to have acted otherwise than as it has done. He is apprehensive that the perverseness of Mr. Seward, the American Secretary of State, who is bitterly hostile to England, and the popular excitement, raised and kept up by some of the newspapers on that side (as it is by the *Times* on this) may bring on a war.

December 31st.

The year closes, leaving the question of peace or war still undecided.

Thus ends 1861. God be thanked for all the many blessings which I and mine have enjoyed during the course of it ; and may He help me to pass the next year better.

Barton.
January 1st. 1862.

I received the new number of the *Natural History Review*, and read Joseph Hooker's article on the "Cedars of Lebanon"; — clear, instructive, and curious. He describes with great precision and clearness the locality of the famous Cedars still existing on the Lebanon. Describes the traces of ancient glaciers descending on the flanks of the mountains to a level of about 6000 feet, s. m.; shews that the Cedars grow only on the moraines of the glaciers; points out the characters which are supposed to distinguish respectively the Cedars of Lebanon, of the Atlas, and of India ; and shews, that the Cedar of Atlas (Cedrus Atlantica) is in the most important characteristics decidedly nearer to the Deodar than to the Lebanon Cedar. Lastly, he shews how, supposing all the three to have originated from one typical stock, inhabiting a region, central with reference to all the three

1862. mountain groups,—changes of level and of climate
(of which the aforesaid moraines are indications)
might have insulated the localities, and gradually
developed in the plants those differences which are
now regarded as specific. This article is, I suppose,
intended to be the forerunner of a larger one on the
same subject. It is well worth studying.

Read in the *Natural History Review*, Mr. Lubbock
on the "Lake Habitations of Switzerland";—
very curious—it deals with a class of facts which
may be said to belong partly to Archæology, and
partly to Geology, or to contain the two. The paper
does not appear to contain the author's own
observations, but a general survey of the researches
and observations of others. It appears that the ab-
original human inhabitants of Switzerland, in times
long before the date of regular History, constructed
their habitations on platforms supported on piles,
in the shallower parts of the lakes, choosing the situations, it is supposed, for the sake of safety, at
first against wild beasts, and in later times against
human enemies.

Great quantities of these piles have been found at
the bottom of the lakes in many localities, and with
them, numerous and various implements, more or
less rudely formed of stone, and horns and bones of
a variety of animals which appear to have been used
for food by the natives. Remains of as many as 66
different kinds of animals have been found; all of
them, with the exception of the Bos primigenius,
still existing on the earth, but several (the Aurocks
or Bison, the Elk, the Red Deer, and the Wild

Boar), no longer natives of Switzerland. But it 1862. would appear that these lake dwellings must have here continued in use for a long time, for while in some localities, all the implements found are of stone, in others they are of bronze, and bear witness to a much more advanced stage of human progress. To say the truth the remark I began with, is not exactly applicable; for these discoveries hardly belong at all to Geology, they are properly Archæological, relating to the early history of *man ;* and connected with Natural History only in so far as they shew that various animals, which are not now wild in Switzerland, were formerly common enough in that country to be used as food.

We had a visit from the Rickards* to-day,—people who I am always glad to see.

Talking of Pitt, Mr. Rickards told us that in his youth he had much acquaintance with Wilberforce, who often talked of Pitt, and described him as a delightful companion. According to Wilberforce's account (and no man had had better opportunities of knowing him), Pitt was full of gaiety, fun, and merriment in private society. This is very different indeed from the idea which all the books give us of him, probably, it was only in the society of his very intimate friends, that he came out in this light.

January 3rd.

Read in the same number of the *Natural History Review*, Mr. Lartet on the Discovery of

----

* The Rev. Samuel Rickards, rector of Stowlangtoft. He used to be called the Keble of the Eastern counties.

1862. Human Remains, associated with those of Extinct Animals in a cavern at Aurignac, in the south of France. This is very curious. The subject is somewhat analogous to that of Lubbock's paper, but is more distinctly connected with Geology and Natural History. I find it difficult to give an abstract of the paper. It appears that at Aurignac, near St. Gaudens, not far from the Pyrenees, there has been discovered in a small insulated ridge of nummulitic limestone, a grotto, or very small cave with a kind of platform of rock before it. Till within the last 10 years, both cave and platform were masked by a talus, or sloping heap of earth and fragments. Accident led to the discovery of the grotto, in which several human skeletons were found. These had been removed long before Lartet's visit, which was in 1860; but even then, in the layer of loose materials which covered the floor of the grotto, he found several human bones, together with flint implements, worked portions of deer's horn, and many mammalian bones. Among the last were remains of several individuals of the great Cave Bear, and Cave Hyæna (both extinct), and a few of the Mammoth. The cave appears to have been a place of burial, and the bones to have belonged to animals killed in the chase by the aborigines, for food or other purposes.

January 6th.

Mr. Tom Taylor, who had arrived on Saturday morning, the 4th, and spent the Saturday and Sunday with us, went away early this morning.

He is engaged in writing the Life of Sir Joshua 1862. Reynolds, and wished to see the pictures by that master which are in this house. He studied them carefully during his stay here, expressed great admiration of them, and gave me some new information as to the dates and circumstances of their painting, as he is in possession of Sir Joshua's original note books. These he shewed us. He is evidently very full of his subject ; is a very good talker, good both in quantity and quality ; evidently a man of very active mind, and of much and various information. He edited so well the lives of Haydon and Leslie that I have little doubt that his life of Reynolds will be as good as the material will possibly admit of.

January 9th.

Dear Katharine and her husband and children left us, to my great regret. I have always great delight in her society ; and her children are charming. This day we received the important news of the American Government having agreed to our demands, and consented to release the prisoners. War is thus averted, at least for the present. As I am not writing political Memoirs of my own time, and as this subject has been discussed in the newspapers till I am perfectly sick of it, I will say no more about it here. This is indeed the principal reason why I have left off putting any politics in this Journal of mine, except when I happen to get any interesting *private* information,

1862. Everything of a public nature nowadays, is not only related in the newspapers, with all its minutest details (true or false), but it is so hackneyed, so vulgarized, worn so utterly threadbare, that I am glad to keep clear of matters of which the newspapers make me so weary.

LETTERS.

Barton,
January 14th, 1862.

My Dear Mr. Horner,

It is rather late in the day to wish you a happy New-year; but, better late than never, and I do most heartily and earnestly wish that the year we have now entered upon, may from beginning to end be a happy and prosperous one to you and yours. We have continued to receive delightfully good accounts of you, and of dear Mrs. Horner and our sisters; and though the climate of Florence seems to have shown itself to you in an unusually unpleasant aspect, it has not, I trust, done harm to the health of any of you.

We have lately had the very great pleasure of visits from the Charles and Henry Lyells; you may imagine how I enjoyed the time. Charles and Mary, indeed, as usual could spare us but few days, but those were very pleasant days. They came to us at Mildenhall, and in the way from Mildenhall hither, Lyell and I stopped to examine a large gravel pit near Icklingham, in which two of the celebrated and mysterious *flint* hatchets had been found. Another day, I took him to see an old

watchmaker at Ixworth, a Mr. Warren, who has 1862.
made a really remarkable collection of antiquities
found in this part of the country;—antiquities
ranging from the '·Stone age," down to that of our
Saxon kings. He possesses the two flint hatchets
found in the gravel pit aforesaid, which Lyell was
anxious to see, as well as many other stone implements of a later date and more skilful workmanship.
I was glad to see Lyell looking well. He was very
intent on the curious questions and researches
connected with the subject of his new book—which,
by the way, he proposed (in joke) to entitle " Pre-
Adamite Man,"—saying that Murray would give
him much more for the copyright with such a title !
I have no doubt that this book will be an interesting
as it will certainly be a very careful and learned and
valuable one.

I need not enlarge on the death of Prince Albert ;
the feeling of sorrow for such a loss and of sympathy
with the poor Queen has been deep and very
general ; I hardly suppose the nation was ever more
unanimous, and it is seldom indeed that it falls to
the lot of a royal personage to be so sincerely
mourned.

The risk of a war with America which seemed
imminent a fortnight ago has passed off for the
present ; whether permanently I feel doubtful.

Much angry feeling has been excited on both
sides, and in the many awkward questions to which
the position of neutrals gives rise, it seems very likely
that fresh *difficulties* (as the Americans call them)
may arise.

1862.   I think our Government has acted extremely well, with moderation, spirit and promptitude, and for once we seemed likely to be ready at the beginning of a war. The spirit of the Canadians appears to be as good as possible. The dispatch of M. Thouvenel and the Prussian one, were very good. It is probable that this affair and the expression of European opinion to which it has given rise will lead to the formation of some clearer and more distinct code of rules as to the rights of belligerents and neutrals.

The Northern States of America really ought to come to some distinct understanding with the rest of the world, as to whether the Southerners are rebels or belligerents ; hitherto they have played fast and loose with the question, calling them sometimes by the one name, sometimes by the other, as suited their purpose. I confess I am not yet at all convinced that the North is likely to conquer the South ; nor am I sure that it is for the interest of the rest of the world that they *should*. I am exceedingly obliged to dear Joanna for her charming letter, for which pray give her my hearty thanks, as well as much love to her and to dear Mrs. Horner and Susan.

Ever your very affectionate Son-in-law,

C. J. F. BUNBURY.

Barton,   1862.
February 5th, 1862.

My Dear Mrs. Horner,

I have two kind letters to thank you for, the last of which came this morning and contained most kind good wishes for my birthday, which I assure you were appreciated. It makes us very happy to hear that you all seem to be spending your time so much to your satisfaction at Florence. You appear to have some very pleasant acquaintances. I admire you and Mr. Horner for your zeal and diligence in studying Italian. I am glad that Mr. Horner has found for himself an interesting occupation in translating the "Life of Savonarola;" I daresay it will be very instructive, and that when I read it—it will open my eyes to the merits of Savonarola, to which as yet I am rather blind. Since I wrote my last letter to Mr. Horner (14th-17th January) we have been very quiet indeed, have gone nowhere, and seen hardly anybody but Henry and Cissy and their children, except when I have gone to Bury on the Petty Sessions days and seen the magistrates. I have hardly anything to tell you. The newspapers will have informed you of the awful disaster in the Hartley Colliery by which 200 men were buried alive, and of the large subscriptions which have been made to help their families. I think this subscription, following so close upon the very large one raised for a memorial to poor Prince Albert, is creditable to England. It seems to be quite undecided yet what form the memorial will take, and of course there is great

1862. difference of opinion ; I think the best plan would be to found some useful institution, for which I do not doubt a sufficient sum could be raised.

(February 8th). I am writing from Mildenhall, we came over the day before yesterday. Snowdrops and winter Aconites are here in profuse blossom, the Mezereon in great beauty, but no Crocuses yet to be seen. Mrs. Marr, Elmer and his wife and child, and the rest here, are flourishing and every thing in perfect order.

In Pettit's stead we have engaged a Scotchman of the name of Allan, recommended by Dr. Hooker and Mr. Smith of Kew. I am sorry for poor Mrs. Byrne,* but I call *him* a very happy man, ending a long, healthy and cheerful life by a painless illness of two days.

Parliament has opened, and I am glad to find so much unanimity in both houses in the discussion of the Address, and such general approbation of the conduct of Ministers on the American question.

I am very glad Professor Parlatore is coming to England this spring, and I hope we shall induce him to come and visit us at Barton. With much love to Mr. Horner, Susan and Joanna,

I am ever,

Your very affectionate Son-in-law,

CHARLES J. F. BUNBURY.

---

* She had lost her husband, Col. Miles Byrne.

Barton   1862.
February 5th, 1862.

My dear Mary,

I thank you much for your kind letter and good wishes on my fifty-third birthday. I have indeed great reason to be thankful for the position in which I find myself, and for the many blessings I enjoy; among which, your and your husband's affectionate and steady friendship, rank very high. I am very sorry for poor Mrs. Twisleton. I feel much for the deep pain it must give you to see one of whom you are so fond, dying so prematurely, and in so much suffering; but above all, I am sorry for her poor husband.* Her fate, and poor Mrs. Bonham Carter's,† are very saddening. Mr. Byrne, I call a very happy man, to have lived to that age with so much cheerfulness and enjoyment of life, and then ended with so short an illness, and with so little pain; but of course, the loss is a sad and heavy one to poor Mrs. Byrne.

I have little or nothing to tell you. The weather is wonderfully mild, as I suppose it is with you also. Snowdrops, Violets, winter Aconites, in full blossom; the Hazel also; and to-day, I have seen Primroses. Little Emily and Henry dined with us yesterday, at our luncheon, and were extremely amusing. They are in great force.

I am carefully studying Heer's great work on the Tertiary Fossil plants, working through it genus by genus, and with great admiration of his sagacity as

* The Hon. Edward Twisleton.

† Daughter of George Nicholson of Waverley Abbey.

1862. well as industry, but I cannot say that I always feel entire confidence in his determinations ; in fact, he himself admits that in some large and important genera he can find *no* certain characters in the leaves, and that his determinations are therefore empirical. I am reading also Lord Clarendon's Life (which I read nearly 30 years ago), and Virgil's Georgics.

Pray give my love to dear Katharine, and many thanks for her kind note which I received yesterday. I hope she will excuse me from writing separately to her just now, as I have absolutely nothing to say.

With much love to Charles Lyell,

I am ever your very affectionate Brother,
CHARLES J. F. BUNBURY.

JOURNAL.

February 7th,

Finished reading to Fanny the Memoirs and Correspondence of Alexis de Tocqueville,—the English translation, edited by Mr. Nassau, senior, and including his notes of conversations with de Tocqueville. These notes are valuable additions to a charming book.

February 13th.

Began to read General Cathcart's narrative of the Russian and German campaigns of 1812 and 1813.

We returned yesterday from Mildenhall, where

we had spent six days. Mrs. Bucke is flourishing. 1862.
Our boys' school was inspected by Mr. Mitchell,
the day before yesterday, and there was a very good
muster. I am very curious to see how the new
Education Code will be discussed in Parliament,
and what will be the result.

February 20th.

We went up by railway to London, to Lillyman"s
hotel.

February 21st.

Attended the anniversary meeting of the Geological Society. Mr. Horner, the outgoing President, being abroad, Murchison, Vice-president, was
in the Chair. Ramsay was elected the new President, Murchison himself gave us an obituary notice
of Dr. Fitton, and did it very well. Then the other
notices of deceased members were read by War-
ington Smythe. Then Huxley gave us, in lieu of
the Presidential Address, a most admirable and
striking discourse on the present state and relations
of palæontology ; beginning with noticing the great
advantage, in all branches of knowledge, of taking
from time to time, a general review of our progress
and of the results actually gained. Applying this to
palæontology, he pointed out that, while our real
advance in knowledge was very great, there was
a tendency to exaggerate the actual gain, and to
place too much reliance on it in certain points.
He applied this particularly to the conclusions
which have been drawn from palæontology as to the

1862. origin of organic life, and as to the successive appearance of higher forms of life. He dwelt much on the insecurity of negative evidence in researches of this sort, ingeniously comparing the case to that of proving by negative evidence the innocence of a prisoner in a court of justice.

He expatiated also on the meaning of geological synchronism, or contemporaneity ; how vague and loose is the sense in which those terms must be understood in geology ; and how cautious we should be not to confound geological contemporaneity, as deduced by organic remains, with contemporaneity in the ordinary sense.

An important and very instructive and interesting part of his discourse was devoted to showing that, great as the differences between the present organic world and that of any remote geological time may appear, the resemblances are much greater and more important.

He mentioned numerous instances of organic types, even of genera which had come down unchanged from the palæozoic times to our own : and showed in detail the comparatively small (surprisingly small) amount of difference, as to orders and larger groups between the organic beings of the present day and those of all previous geological periods.—On the whole, I have rarely listened to a scientific discourse more calculated to suggest thought and enquiry, to excite the faculties or to provoke controversy.

Anniversary dinner of the Geological Society at Willis' rooms :—I sat by Charles Lyell. The Duke of Argyll and Monckton Milnes were present as 1862. guests, and both spoke—and spoke well.

Met Joseph Hooker at the Athenæum, and had much talk with him. He told me of a most extraordinary new plant, of which information has been received (and as I think he said specimens also)— discovered by Welwitsch in Southern Tropical Africa, south of Benguela, and nearly at the same time by another traveller in the Dammara country : a plant more extraordinary, he says, both in appearance and in essential peculiarities of structure than Rafflesia, or any other plant hitherto known. He told me also, that the enterprise in which he and all the Kew establishment have had a great concern, of introducing the most valuable kinds of Cinchona* in a living state from the Andes into British India, has been successful; and that several hundreds of young Cinchona plants are now growing on the Neilgherries and the mountains of Ceylon. This appears to be an important achievement ; not only with the very large sum which India has hitherto had to pay annually for Cinchona bark be saved or reduced, but the world in general will no longer be dependent for its supplies of Quinine on the *wild* growth of a few limited districts in the Andes. The Dutch, indeed, had already successfully introduced the culture of the Cinchona into

* Introduced by Mr. Clements Markham.

1862. Java. Hooker told me that his paper on the Cedars in the *Natural History Review*, was, (as I had understood) intended as a prelude to a larger and more elaborate memoir on the subject, for which he is collecting materials. One of the oldest and largest Cedars on Lebanon, has lately been blown down in a storm, and Hooker is anxious that it should be transported to England and deposited at Kew.

Huxley joined us, and we had a good deal of talk on various points connected with the subject of his discourse the day before. Hooker told me that he had attended a lecture at the Royal Institution, by Mr. Ferguson, the famous architect, on the Holy Sepulchre at Jerusalem. In this, Mr. Ferguson attacked the generally received tradition as to the site of the original church of the Holy Sepulchre ; and Huxley, who has himself seen the localities, thinks that he has completely made out his case.

In talking of Geological synchronism, and the wide sense in which the word ''contemporaneous '' must be understood, I made the remark (as the other two assented) that we, who are now living, are in a *geological* sense contemporaneous with the Mammoth and the Siberian Rhinoceros, since they have been proved to have co-existed with man.

February 27th.

With Fanny and Lord Arthur Hervey, to the British Museum. Saw Mr. Charles Newton, who was very attentive in shewing us its great collection of antiquities which he had brought from Halicarnassus and Cnidus. These are the subjects

of the magnificent book he has lately published. 1861.
They are still shamefully ill-housed, or rather stowed
away, in mere temporary boarded corridors; but we
saw them to great advantage under his guidance,
and he was very pleasant.

March 11th.

Charles Lyell tells me that he has seen the
specimens which are at Kew, of that wonderful
plant from Africa, the Welwitschia. He says that
it is as uncouth and strange in its outward appearance, as it is anomalous in structure. Hooker
considers it more anomalous and extraordinary,
more irreconcilable with hitherto admitted systems,
than any of the extinct plants of the Coal formation.
As I understand the accounts, the stem, though
woody, never increases in height, but keeps expanding laterally, so as to form something like a
great plate or cake of wood on the surface of the
ground; it has never more than two leaves, and
these are in fact its cotyledons, which are permanent,
and continue increasing in size as the stem goes on
expanding; the flowers come out round the edges of
the disc-like stem; and besides many other
peculiarities, the structure is *gymnospermous*, the
ovules being entirely exposed; yet there is nothing
else in common with the gymnospermous plants
before known.

Lyell has also told me of a botanical discovery
lately made near Boston, in New England, which
has excited great surprise and attention among the

1862. American botanists,—the discovery of our common
heather (Erica or Calluna vulgaris) growing
abundantly in one spot, a short distance from the
city of Boston. The entire absence of all heaths
from America, has always been supposed to be one of
the best established facts in botanical geography.
It seems very extraordinary that such a well known
and easily recognised plant as the heather, should
have escaped observation in the neighbourhood of
Boston and of Cambridge University; yet, it is
difficult to suppose it introduced, the plant being
notoriously difficult to cultivate or to introduce when
it does not grow naturally.

March 15th.

We came down to Barton.

March 16th.

I read part of Joseph Hooker's paper on Arctic
plants, in the new part of the Linnean Society's
Transactions; and resumed the reading of General
Cathcart's narrative of the campaigns in Russia
and Germany in 1812-1813, which I had begun in
February, and left unfinished when we went to
London.

March 17th.

Finished General Cathcart's book. It is a very
clear and distinct, and (as appears to me) sensible,
but not lively narrative of the events of those two
great and extraordinary campaigns which broke the
power of Napoleon. The writer (afterwards Sir

George Cathcart, who was killed at Inkerman) was a spectator of many of the battles, being on the staff of his father, who was in attendance on the Emperor of Russia, as a sort of military diplomatic representative of Britain ; yet, the narrative has little of the vividness of personal impressions. It is only on one or two occasions that he specially mentions circumstances as having come under his own observation ; as, when Napoleon was distinctly seen by the allied sovereigns and their suite, just before the battle of Bautzen, in close conference with a supposed officer in a yellow uniform, who excited much curiosity and speculation, but who turned out to be a Saxon postillion ;—and again, in describing the entry of the Allies into Leipzig, and the dreadful scene exhibited by the village of Probstheide, which had been so desperately con-tested throughout the preceding day. I also finished reading Hooker's very elaborate paper on the distribution of Arctic plants, in the Linnean Society's Transactions,—like all his writings, most accurate, complete, and masterly.

March 18th.

To Ipswich with Fanny, in our own carriage, returning at night. Attended a great county meeting, called to consider the proposal of establishing a great Middle-class School and Agri-cultural College for Suffolk. Lord Stradbroke presided. The meeting altogether, was very harmonious. All that was proposed was very well received, and several large subscriptions

1862. started the scheme with great *éclat*; but the difficulties will begin when we begin to reduce it into practical shape.

The Assizes at Bury. I was foreman of the Grand Jury, and dined with the Judges; sat next to the Chief Baron Pollock, who talked much to me; —a fine, spirited, animated, active-minded old man, who told me himself, that he would be eighty years old next September.

Began to read Villemain's " Souvenirs Contemporains."

Up to London again.

We visited Sir Frederick and Lady Grey—much talk with them about the rival Codes of Education— they are both very keen for the new code (Mr. Lowe's).

The. first news of the remarkable naval engagement in the James river, in North America,—the victory of the iron-plated frigate, Merrimac, over two wooden frigates, the Cumberland and Congress, in the drawn battle between her and another iron ship, the Monitor;—a practical experiment which

seems to establish conclusively the tremendous 1862.
superiority of iron ships, and the helplessness
against them, of the best timber-built ones.

To the Linnean Society; studied there for some
time. Kippest showed me Weddell's fine work on
the Cinchonas, with its beautiful plates (by Alfred
Riocreux) : some of the most exquisite specimens
of botanical drawing that I have seen. The figures
of the plants are uncoloured, and at first sight
appear little more than outlines ; but all the textures
of the different parts,—stems, leaves, flowers, and
fruits,—are rendered with a delicacy and truth of
effect, beyond what I have seen in almost any book.
The plates of the barks are coloured, and represent
their various textures wonderfully well.

Visited the Zoological Gardens ; looked more
especially at the birds. Noticed two fine specimens
of the Jungle Cock, Gallus sonneratii ; handsome
birds, lively and noisy ; their *crow* is very much
harsher, more broken and irregular than that of the
domestic cock ; one might call it a poor imitation
of a genuine cock crow. In the Parrot house, six
fine live Toucans, of three different species, all
healthy and lively ; one (from Mexico) belongs to
the Ramphastos carinatus, figured by Swainson,
and is very remarkable for the brilliant and strangely
variegated colours of the beak, even beyond what
is shown in Swainson's plate. The other two

1862. species, Ramphastos ariel, and Ramphastos toco, are Brasilian. The mode of feeding, in all of them, is peculiar ; they take up the food with the point of their monstrous beak, and throw their head back, exactly as if they were swallowing a pill. The appearance of this enormous beak, in the living, as well as in the stuffed bird, is certainly very strange and grotesque—like a mask. Though the Toucans have the toes placed as in the Scansores, they hop, and do not climb. The birds of the Parrot family, in this house, are amazingly numerous, amazingly beautiful, and inconceivably noisy.

In the Reptile house I saw the famous female Python brooding over her eggs, but did not see the eggs themselves.

N.B.—Her incubation has since proved a failure.

———— ——

March 30th.

We went to Vere Street Chapel, and heard a fine sermon from Mr. Maurice,* on St. Paul at Lystra ; —fine observations on complying with popular superstition for a supposed good end, and on intolerance and persecution.

Had luncheon with the Henry Bruces, and a pleasant chat with them. Henry Bruce,† very pleasant ; thought the Government very wise in compromising the question of the Education Codes; but thinks much better of Mr. Lowe than I have been accustomed to do ; (N.B.—He knows Lowe intimately, while I have no personal acquaintance

* Rev. Frederick Maurice.
† Now Lord Aberdare.

with him). He believes Mr. Lowe is in earnest in
his political opinions, but says he is a "doctrinaire,"
and has no tact or judgment as to what can, or
cannot be carried into effect; and that even his
extreme shortness of sight is a serious disadvantage
to him and the House, as it prevents him from
observing men's countenances, and seeing how what
he says is received. He quite agreed with me that
the tone and manner in which the "Revised Code"
had been advocated in the *Times* were intolerably
offensive, and had had a great share in exciting
animosity against it.

April 1st.

Went with Fanny and Katharine to Veitch's
nursery garden in the King's road, Chelsea ; — a
very great establishment, with an immense extent
of glass, and a noble collection of plants, of which
we saw only a part. Ferns in great variety and
beauty,—several rare ;—of Schizœa dichotoma, the
only plant yet in Britain ; some others, supposed to
be entirely new, and not yet named. I was
particularly pleased with Lycopodium Phlegmaria,
which I never before saw alive. Several species
of Gleichenia flabellata, dichotoma, dicarpa, and
microphylla, in great beauty. Some Cheilanthes
of exquisite delicacy. Several fine Tree Ferns
(Dicksonia and Alsophila) which were brought out in
the state of *logs*, now putting forth vigorous and
beautiful fronds. We noticed also some beautiful
Orchids, and flourishing young plants of Dionæa
Muscipula, and Cephalotus follicularis.

1862.   We dined with the Henry Lyells ;— the party : —
Lady Bell, Sir Edward Ryan, the Woronzow,
Greigs, Erasmus Darwin, Dr. Falconer and his
niece. In the evening, Mr. Maurice, and two
Miss Sterlings, Mary, Sally, and Louisa, the
Charles Mallets, and several more. Altogether, a
very pleasant party.

Sir Edward Ryan's very high opinion (to which
I entirely subscribe) of Lord Canning; thinks he
has done more than any other man to raise the
condition of the natives of India, and that he
deserves honour for the firmness with which he held
the balance between the natives and the Europeans
during the Mutiny.

Greig's description to me, of Ricasoli's old Castle,
in the Apennines; quite like a castle in Mrs.
Redcliffe's romances.

April 2nd.

Geological Society :—the Club. Sat by Douglas
Galton and General Tremenheere, the latter, an
Indian officer, who gave me a striking description of
the tropical forests of Tenasserim, where he had for
some time the care of the Teak trees.

Talk about the iron ships, and about the unsatisfactory answers given to enquiries on the
subject in the House of Commons by Sir George
Cornwall Lewis and Lord Clarence Paget — Sir
George Lewis's in particular, a thorough red tape
speech. Douglas Galton thought that our Government would never learn from anything but actual
disaster. (It has since turned out however, that

the Government were more awake to the importance 1862. of the subject, than the aforesaid speeches seem to imply). Douglas Galton disapproves of the plan of the fortifications of Portsmouth and thinks an elaborate chain of forts along the whole length of Portsdown hill, superfluous.

April 3rd.

A very fine morning. With Henry Bruce and Fanny to Little Holland House, that charming place which was so familiar to me when my dear old friend Miss Fox lived there, now inhabited by Mr. Prinsep, a noted oriental scholar, whose wife was a Miss Pattle, sister to the beautiful Lady Somers. We went to see the paintings of Mr. Watts, who has a studio there, and we were very much pleased. They are portraits and ideal subjects: the latter showing a fine poetical imagination and elevated taste; the style of execution reminded us of some of the Italian frescoes. Mr. Watts himself is an interesting man. I am much struck by Henry Bruce's variety of knowledge, and the activity, clearness and vigour of his intellect.

Meeting of the Linnean Society. A very remarkable paper by Charles Darwin on that curious anomaly in Orchids (first noticed by Sir Robert Schomburgk in Demerara, and afterwards by Lindley and others) of the occurrence of flowers of supposed distinct genera on the same plant and even in the same spike. He took up Schomburgk's instance, of which the original specimens have been preserved

1862. in the Linnean Society, where flowers of so-called Catasetum tridentatum, Monachanthus viridis and Myanthus barbatus occur together. Schomburgk *suspected* that the differences were sexual; Darwin, by a most minute, elaborate and sagacious examination, proves that this is the case—that the Catasetum is the *male* flower, the Monachanthus the *female*, and the Myanthus the *hermaphrodite*. *These* usually occurring on separate plants, had very naturally (as they are extremely unlike) been taken for distinct species and even genera : but now and then flowers of two and even three kinds are produced on the same plant. This is, I believe, the first ascertained instance of separated sexes in Orchids; it is very possible that in other instances also, different sexes of the same plant may have been taken for different species. Darwin gave us part of his paper *vivâ voce*,—a beautiful exposition of the curious and complicated structure of the sexual apparatus of Catasetum. Joseph Hooker gave him due honour in the few remarks which he made upon the paper, and observed at the same time that the whole subject of Orchidaceæ had for a long time past been in a manner given up by professed botanists to two men, Brown and Lindley. The discussion which followed was for the most part *beside* the matter of the paper, going off into the question of vegetable irritability of which there are some remarkable examples in Orchidaceæ.

We went down to Mr. Sam Smith's, Combe Hurst, and spent the afternoon very pleasantly with him and his two daughters, Mrs. Clough and Mrs. Coltman : charming people all of them. He is one of the pleasantest men I know. The weather was miserable, but we enjoyed the beautiful place as much as we could under the circumstances.

<hr>

Went to the British Museum, and spent some time in the botanical room, where the fine specimens of Tree Ferns and Palms are, and the specimens of curious vegetable structure ; but I did not observe that anything remarkable had been added to the collection here, since I saw it in March, 1861.

Within the last week fresh experiments at Shoeburyness with the Armstrong gun, have shown that by employing a very heavy charge of powder, so as to give the shot a very great velocity at starting, any iron armour hitherto used for ships can be utterly smashed. It is a curious and exciting race that is going on between the improvements in artillery and in the construction of iron ships ; each by turns appearing to gain the ascendancy. It is a tremendously expensive contest, but the satisfactory determination of the question is of such vital import-ance to England as a nation, that we must make up our minds to the expense. I think it most likely that in the end the guns will win, for it would seem that there must be a limit to the thickness of the iron casing which ships can carry if they are to float.

1862.   April 8th.

We went to see Mr. Tom Baring's pictures; a beautiful collection in a beautiful gallery. The weather wretched.

April 10th.

Sat to Boxall; the seventeenth sitting, and I hope the last. He began my picture while staying with us at Barton last summer. The weather miserable during the last three weeks. Nothing but drizzle and fog; the sky coloured like the reflection of the mud of the streets.

April 11th.

We travelled post down to Barton.

LETTERS.

Barton,
May 26th, 1862.

My Dear Katharine,

I must write you a few lines, to express however imperfectly, my sincere sympathy with you in the great grief which has fallen upon you all, as well as my own sorrow for the loss we have all alike sustained. I did indeed love dear Mrs. Horner very much, and good reason I had, for nothing could possibly surpass her constant kindness and indulgence to me; she could hardly have shown me more affection if I had been her own son.

It is a great comfort, and I think after a time you and Fanny and all of you will feel it to be a comfort

to reflect how bright and happy her life was; how 1862.
prosperous and how blessed with that disposition
(so far above all worldly gifts) which Addison so
beautifully speaks of that "cheerful heart which
tastes those gifts with joy." How happy she was
in her family! and how she preserved not merely
her faculties, but the full zest of life and capacity of
enjoyment, to a mature old age, and almost to the
very brink of the grave. Truly I hardly see how
any mortal could have a happier lot than that
which the goodness of God granted to your dear
mother. And we must trust to that goodness to
soothe and heal the sorrowing hearts of her children,
and still more of dear Mr. Horner. His loss is by
far the heaviest, and for him I feel the most; I am
very anxious to hear how his health bears this
terrible blow.

Fanny is still much overcome, and not yet able to
rally from the shock; it is better that her grief
should have its free course, but I am rather afraid of
its hurting her health ; I am glad to say she is under
Image's care.

Your beautiful and interesting children, and the
attention they require will be *your* best consolation,
dear Katharine. It would be a great pleasure to us
to see you here, but I can well understand that at
such a time you would rather not leave your home.
But you will also be able to understand that
for Fanny also home is best.

With much love to your husband and children,

I am ever your truly affectionate,

C. J. F. BUNBURY.

Barton,
September 14th, 1862.

My Dear Katharine,

Many thanks for your very kind and pleasant
letter of the 9th.   I am very glad you liked the
Ferns.   I felt quite ashamed when it was too late to
think how few I had sent you.

I was very sorry to part with your sisters, and
with the dear children ; these had come to be on
excellent terms with me, having quite got rid of
their original shyness, not that I think Dora was
ever very much troubled with it.   She is the
merriest little mortal.   I miss their merry little
voices and bright faces very much.   I trust Mr.
Horner gained somewhat in health by his stay here ;
I certainly thought he looked better, and in
particular walked more stoutly latterly, than when
he first came.

We have had splendid weather for the harvest,
which is now almost completely finished in this
parish, and indeed I believe in the neighbourhood
generally ; and we have great reason to be thankful,
for the wheat crop though not large in quantity, has
been gathered in in the finest condition, and most of
the other crops are very good.   A wet harvest, with
all the other circumstances of this present season,
would have been terrible.—The mornings have been
particularly beautiful for some time past ; I have
been out for half-an-hour before breakfast every
morning, strolling about the lawn and garden and
arboretum, and have enjoyed it exceedingly.   The
exquisite clearness and freshness of the air, the

exhilarating brilliancy of the morning sunshine, the 1862. flowers and grass sparkling with dew, the hum of innumerable insects about the flowers, and the cooing of the woodpigeons, make these autumn mornings very delightful. The days, though free from rain, have not as a general rule, been fine in proportion. Our Ferns are flourishing, and I shall have some new ones to shew you when you next come here; in particular, two exquisite little species of Cheilanthes: one Cheilanthes elegans, is like fine filigree, or like *green frost work*, if one can imagine such a thing. The other, also very delicate, we had from Veitch, under the name—certainly wrong, of Cheilanthes glauca var. hirsuta (N. B., it is neither glaucous nor hairy!)—I have not yet made out what it is.

Fanny and I have been very busy lately, planting (or rather choosing the places for planting) a number of new exotic Conifers and other trees and shrubs, which have arrived from Veitch's. It is a pleasant occupation.

We also go on steadily with the library catalogue, which is very entertaining work, and makes me far better acquainted with the books we have, than I was before. I have read wofully little during this last month, but have large schemes of reading, which it is sadly difficult to keep to, as so many new books come in the way, and one is expected to read them.

I was very glad to hear of what you had been doing at Freshwater. I was sure it would be an interesting place, both for you and Leonard. For

1862. geology, indeed, one can hardly find a more interesting place ; what with the fine cliffs of flinty chalk on each side of Freshwater, the Tertiary strata of Alum Bay on the one side, and the green sands and Wealden beds of Compton Bay, and Brook Point on the other. But you don ot seem to have found many plants characteristic of the island, or of the south of England ; at least, all those you mention, except perhaps the Centaurea Calcitrapa, grow at Mildenhall.

I am very, *very* sorry for Garibaldi ; sorry that he should have plunged into a scheme of such culpable rashness, almost madness ; even more sorry, if possible, that he should be wounded and a prisoner. I have as little doubt as ever of his honesty of purpose, and think it very possible that he has been entrapped. M. la Guervoniere's declaration of French views regarding the unity of Italy, is instructive at any rate. It shows what was the "idea" for which France went to war in '59 ; and how much faith Italy ought to place in the Imperial friendship.

Ever your very affectionate Brother,
CHARLES J. F. BUNBURY.

Barton,
October 10th, 1862.

My Dear Mr. Horner,
        We returned the day before yesterday from Cambridge where we had spent eight days very pleasantly.
        The British Association meeting was a very

pleasant one, and I believe a very good one in a 1862. scientific view, but not so good in a *money* view, at least so I heard.

Your absence, and Charles Lyell's were much regretted, and many inquiries were made after you ; Murchison also was missed ; but Sedgwick was in great force. It was a great pleasure to us to see much of him, and of the Kingsleys.

Jukes, as President of the Geological Section, opened the business of it with a capital address, or lecture, treating of the causes which have produced the present form of the Earth's surface ; and especially of *denudation*, to which he ascribed an altogether predominant share in producing and modifying that form. Indeed, he carried this doctrine very far, claiming for the denuding power the whole work of modelling the surface with the exception only of actual volcanic mountains ; contending that the upheaving and disturbing forces, however great, have no other effect *on the surface* than as they bring strata of various hardness and consistency within reachof the denuding agency. In short, Denudation seemed to be his hobby, and he rode it pretty hard. It is not etiquette to make any reply to a Presidential Address, otherwise, there was much in Juke's that might have invited controversy ; in particular, he gave his sanction to Ramsay's daring hypothesis of the glacial origin of Lakes. The worst of these Association meetings is, that one cannot well be in two or three different Sections at the same time, and consequently, one is sure to miss something. Thus, while I was listening

1862. to Juke's introductory address, which was well worth
hearing, I lost Huxley's, which, I understand, was
still better ; however, it is to be printed, and he has
promised me a copy.   I missed also, the great
battle between Huxley and Owen, the next morning,
which excited immense interest ; that is, I could not
get near enough to hear it ; the room being excessively crowded.   The general impression seems
to be, that Owen had the worst of it.   I am really
sorry that with all his just claims to veneration, he
will put himself in the questionable positions, and
expose himself to the risk of a public defeat.
There were several good papers in the Geological
Section :—a good account by Daubeny, of the last
eruption of Vesuvius, with observations on the
peculiarity of its gaseous products, and their
differences from those before observed ; remarking
on the importance of attending to chemical facts
and laws in reasoning on volcanic phenomena.   A
paper by Phillips, on Slaty Cleavage ; very well
delivered (spoken, not read), and seemingly as
intelligible as the nature of the case admitted ; but
it is to me one of the most difficult parts of geology,
and one of those of which I know least.   So far
however, I made out that Phillips agreed with Mr.
Sorby in attributing the phenomena of slaty cleavage
entirely to mechanical pressure ; whereas, Sedgwick
(who spoke with much force and animation on the
subject), thought that they could not be sufficiently
accounted for by that kind of agency, but were
owing to a crystalline force.

Falconer gave us a clear and curious account of

the Bone caves discovered in Malta by Captain 1862.
Spratt, and the discoveries that he (Falconer) had
made among the bones from them ; in particular of
the pigmy Elephant, no bigger than a dog, that formerly inhabited what is now Malta.   There was also a
curious paper by a Mr. Moore, describing certain
veins in the carboniferous limestone of Somerset-
shire, which are filled up with limestone of a later
date, wherein he had discovered fossils of Oolitic
age, and in great abundance ; also giving an account
of his discovery of organic remains in the clay contained in certain metalliferous mineral veins of the
North of England.   His researches seem to have
been carried on with remarkable energy, and on a
large scale; for instance, from some veins in a quarry
near Frome, he had had a quantity of matter
carted 20 miles to his house, had spent two years in
examining it, and had obtained from it 50,000 teeth
of one species of fish, which he showed us.   (N.B.—
We did not count them.)   He had obtained also,
from another locality, as many (I think) as 25
specimens of *Microlestes*, the most ancient mammal
as yet known.   This paper led to a rather good
discussion on mineral veins, in which Ansted took a
prominent part.   It seems impossible to escape the
conclusion at which he, as well as Mr. Moore,
arrived ; that the organic remains found in mineral
veins must have been introduced from above, and
therefore, that such veins must at one time have
been open to the surface.

Of course, you will by-and-bye find all these
things much more fully set forth in the Transactions

1862. of the Association; I merely give you a few sketches of what most excited my attention.

Many thanks for your information concerning the work on Mineralogy; but if it be, as its title implies, arranged in the manner of a dictionary, I am afraid it will not answer my purpose. What I want is a *system*, tolerably clear and intelligible, by which I may arrange the collection formed by my father.

I hope you will soon find yourself comfortably settled and at home * in your new house, of which I am glad to hear a very good account.

With much love to Susan and Joanna,

I am ever your affectionate Son-in-law,

CHARLES J. F. BUNBURY.

Barton.
October 11th, 1862.

My Dear Katharine,

If you look into the Athenæum (the newspaper I mean, not the Club) you will find a very dismal account of the British Association meeting at Cambridge : but my own impression of it is very different from that author's—except only as to the badness and dearness of Cambridge inns. To me it was a very pleasant meeting, rendered especially so by the Kingsleys and Sedgwick, though several were missing of those whom I remember playing a distinguished part at the few former meetings of the kind which I had attended, some staying away as your Father, Charles Lyell and Murchison, some

* In Montague Square.

gone to a better world, as poor Edward Forbes and
Henslow.

Dear old Sedgwick was in prodigious force,
looking remarkably well; I have not known him
more in his glory for many years than he was this
time,—and indeed also when he was with us here in
the spring. His cordiality and kindness to us were
most marked. He spoke beautifully at the great
dinner which was given in the hall of Trinity on the
Friday, and with great spirit, more than once in the
Sections. Indeed for a man of 77, his speaking
appeared to me wonderful. It was a great pleasure
to me to meet the Kingsleys again. Kingsley
himself, indeed, was so busy that I had not as much
talk with him as I should have wished, but his
wife I like more and more in proportion as I know
her better, and Rose is a perfectly charming girl:
quite a model of a girl. I had much enjoyment in
our visit to Cambridge, but there was nothing more
pleasant to me than our last quiet evening with the
Kingsleys.

There seem not to have been any very important
botanical papers in section D; much the greater
number of those that were read there were on
Zoology. But on the Tuesday Mr. Symonds read a
short notice of a peculiar locality of Asplenium
viride on the Black Mountain in Monmouthshire
where he found it confined to a limited deposit of
traverstine in the midst of a red sandstone district.
This gave Kingsley and me an opportunity of
making some remarks on the distribution of Ferns.
Next to this came a paper on the botany of a very

1862. peculiar limestone district in the county of Clare in
the West of Ireland ; a district noted for Gentiana
verna, Dryas octopetala, Adiantum capillis veneris
of splendid growth, and other rarities.   A great
part of this country, it seems, consists of a sort of
natural pavement of limestone with deep crevices in
which the plants I have named, and many others
grow with extraordinary luxuriance ; while on the
surface of the horizontal slabs of limestone there is
hardly any vegetation at all.   One of the few other
papers that I heard in that section, was a curious
account by Mr. Lubbock, of an *aquatic wasp* that he
has discovered: an insect of the wasp kind that
lives entirely in the water, has well formed wings
but uses them only for swimming, never for flying.
A very odd arrangement, don't you think so ?   So
much for the learned.

William Napier and Edward are here, both very
flourishing ; but we have been disappointed of the
Maurices and several other invited guests.

I hope we shall be in town for a few days at the
end of this month, and that we shall then see you
and your dear children.   Give my love to them all,
and also to Harry Lyell.

Ever your very affectionate Brother,

CHARLES J. F. BUNBURY.

————

Barton,
January 14th, 1863.

My Dear Katharine,

We have been very quiet since your dear
1863.   Father and Sisters left us, and the time has seemed
to pass wonderfully fast.   I am often astonished at

the speed with which the days and the weeks flit by —the morning comes, and almost before one has time to look about one, seemingly, it is evening again. I never find any day half long enough for what I want to do. I may truly say as the children say, that I am "as happy as the day is long," but the days seem very short! The only deviations from "the even tenour of our way" have been—a journey to Ipswich on the 6th, where I had to attend a meeting of the Committee of the Albert College— attendance yesterday morning at the Quarter Sessions at Bury, where the adoption of the new Highway Act was discussed and decided upon—and a pleasant dinner party yesterday evening at Mr. Beckford Bevan's. We met there—the Arthur Herveys—whom I am always glad to meet, and some other pleasant people.

Our happy seclusion will be broken in upon, in a few days by Cecil, who is to spend a fortnight with us ; and before long probably by Lady Bunbury, but her movements are most uncertain.

January 27th, 1863.

The accounts of poor Lady Bunbury are very sad indeed ; there seems every probability that she is really dying ; very little chance that she can really recover ; it is a painful and trying time for poor dear Cissy.

I have not (on account of a bad cold), been able for a long time to look at my dearly beloved Ferns, but Fanny reports well of them — the living ones I mean. The conservatory is gay with Chinese Primroses, Yellow Jasmine, Justicia,

1863. Sericographis, Cyclamen, Scarlet Euphorbia, Coronilla and the bright berries of Solanum capsicastrum.

We have a very extraordinary winter ; no snow at all, and no sharp cold, and not much rain, but an almost continual succession of gales of wind. Scott says he hardly ever saw the park so dry at this season.

I have lately finished " Stanley's Palestine ; " the clearest and most complete and interesting account of that country that I have read ; before this I had read " Wallace's Travels on the Amazon and Rio Negro; " a very satisfactory book of travels indeed ; full of information, in a clear, simple and unaffected style. I am also reading " Thiers' History of the Empire." Pertz says it is full of lies, and it certainly is thoroughly one-sided; but very well written and very pleasant reading. I have read with much satisfaction Dr. Falconer's paper on Elephants in *The Natural History Review.*

With much love to all your family, great and small,

I am ever,
Your very affectionate Brother,
CHARLES J. F. BUNBURY.

Barton
February 4th, 1863.

My dear Mary,

Very many thanks for your kind and pleasant letter. I am very much obliged to Charles Lyell for his book, which I am delighted to see really *out* at last, and which I look forward to reading with great pleasure and interest. I have 1863. already, since it came, read the first two chapters ; the curious history of the Danish Refuse-mounds, and the Swiss Lake-dwellings, I knew already from Mr. Lubbock's papers, but it is very clearly and forcibly told in Lyell's second chapter. Indeed, it is evident that the book is as full of matter as it can hold, and I do not wonder that it should have cost him so much time and labour. I have no doubt that it will add much to his well earned fame. I hope, now it is fairly out, that he will take a good spell of rest and relaxation.

It is true, as you say, that many books of importance, or of some interest to us for family reasons, are coming out just now. But the worst of all these new books (if one reads them) is, that one's reading becomes altogether desultory and disjointed. This does not apply to Lyell's book, which comes into my regular course of scientific reading ; but it is sadly difficult to keep to any system, or to find time for reading older books.

My Fanny has to-day made me a present of " The Prince Consort's Farms" ; have you seen the book ?—it is very handsomely got up, and is highly spoken of.

Of poor Lady Bunbury I hardly know what to say ; almost every day's account contradicts that of the day before, except in so far as they are all lamentable.

Lord Lansdowne may well be said to have been a happy man, to die so easily, after such a long, prosperous, and honourable life. One cannot feel

1863. anything like pity for him; but, as you say, he formed a remarkable link with the past. To think that he was Chancellor of the Exchequer under Fox;—that he succeeded Pitt in the representation of Cambridge;—and moved the vote for the cost of Nelson's funeral. It seems to take one back almost to patriarchal times. I thought the biographical sketch of him in the *Times* very good.

My cold and cough are much better. Fanny persuaded me, to-day, to take a drive with her, and to walk a little on Rougham Heath, and it did me no harm. She is not quite well herself: indeed, I think this strange, unnatural season agrees with hardly any one. We go on regularly with the Library Catalogue.

With much love to Lyell,

> I am, ever your very affectionate Brother,
>
> C. J. F. BUNBURY.

Barton,
February 7th, 1863.

My Dear Katharine,

I thank you very much for your kind letter, as well as for the pretty little pencil-case which you sent me on my birthday, and pray give my love to darling little Rosamond, and many thanks for her pretty present. As I wrote to you lately, and have since written to Mary, and to Mr. Horner, I have little new to say, except, that Henry has come back from Wales for a time, poor Lady Bunbury being still seriously ill. He evidently was much in need of a change. William Napier and Sarah Craig are

with her now, and Henry means to go back when
William has to return to Sandhurst.

(February 9th.) The account from Wales this
morning is better ; but, she believes herself dying.
Poor Cecilia is going to set off to-morrow to see her.
Dear little George seems very materially better, and
we may venture to hope that he will in time recover
completely. My cough is better, and I am gradually
gaining strength. We are going to Mildenhall
to-morrow, though I feel very much disinclined to
move, but Fanny has set her heart on it, and thinks
it will do me good.

I am reading Charles Lyell's book with great
interest and satisfaction. I have just finished the
twelfth chapter. I have got the new part of the
species Filicum, but have as yet only glanced at it.
I have great doubts whether I shall adopt his views
as to Polypodium, which he takes (I see) in as com-
prehensive a sense as that in which it was understood by Swartz and Wildenow.

The season is very extraordinary, the Honey-
suckles and some of the Roses are quite in leaf ;
and the following plants are in blossom in the open
ground about our house :—Yellow Crocus, Purple
Crocus, Sweet Violet (in profusion), Crimean and
Common Snowdrops, Cynoglossum, Omphalodes,
Wallflower, Primrose, Anemone coronaria, Scilla
Sibirica, Spurge Laurel, Laurustinus, and Arbutus ;
besides two wild flowers, Ranunculus Ficaria, and
the Daisy.

There will be a sad deal of mischief done
when the North-east winds come, as they are sure to

1863. do, and I daresay we shall have bitter weather in April and May.

I was very glad to hear of dear Leonard having such a pleasant expedition with his uncle and aunt. Give my love to him, and to the rest of your children, as well as to Harry.

Ever your very affectionate Brother,

CHARLES J. F. BUNBURY.

[In the middle of February, Sir Charles Bunbury and his wife went to London, to furnish the house 48, Eaton Place, of which they had bought a long lease. Lady Bunbury died about the middle of March, and they returned to Barton to attend her funeral. She had been an excellent and affectionate step-mother, sympathising in all his joys and troubles, and he was much attached to her. She had a brilliant intellect, and a noble, generous heart, and ever since her marriage, had been the greatest comfort to his Father, and to Sir Charles, and his brothers. After his Father's death in 1860, she spent nearly three months with him at Barton, and paid him long visits in 1861 and 1862, when he much enjoyed her society; he used to read to her in the evenings, and her remarks on books and politics were most interesting. In May, 1862, when she was at Barton, they heard of the sudden death of Mrs. Horner, and Lady Bunbury's warm sympathy in this overwhelming sorrow, was a great comfort to them both.]

My Dear Katharine,

The Trichomanes radicans gathered by Mr. Fox in the island of St. Sebastian, Brazil, is unquestionably the true genuine Trichomanes radicans. The type of the species is a West Indian Fern, of which I have seen no specimens, but by the descriptions it does not seem to differ very much from the Brazilian. The Trichomanes speciosum, Willd.—native of the Atlantic islands—is united by Sir William Hooker to Trichomanes radicans; it certainly looks very different indeed, at least from the Brazilian form; but as I said, I have no means of comparing it with the West Indian plant on which the species was founded, and Sir William says there are intermediate forms. Indeed I see a considerable difference between my Madeira and Teneriffe specimens of Trichomanes speciosum; and Joseph Hooker on seeing my Teneriffe specimens, was struck with their great difference from the Irish Trichomanes, which Irish plant is now identified by all authors with speciosum. The whole subject is very puzzling.

I have never yet thanked you for the parcel of Ferns you sent me just before we left home. They are welcome and valuable additions to my collections, especially the Scolopendrium, Hemionitis and the Marsilea. This Scolopendrium seems to have been much confused as to its history and synonyms, with the Asplenium palmatum, from which it is in fact widely different. The confusion,

1863. it would appear, began with Linnaeus himself, who
gave the name of Asplenium *Hemionitis* to the
palmatum, taking it (seemingly) for the *Hemionitis*
of the old herbalists : whereas their Hemionitis
seems to have been really the Scolopendrium, which
is an Italian plant.   I remember noticing among
your Ferns from Professor Parlatore, a Cystopteris
Taygetenus, with the synonym Aspidium Tayg. of
Bory.   I found this next day in Bory de St.
Vincent's description of the Cryptogams of the
Morea.   (Exped. Scintif. de Morèe.)   He found it in
holes and clefts of rocks, in the highest parts of
Taygetus, 1800 or 1900 metres above the level of
the sea.   He says, it has a great resemblance to
those delicate Alpine species which are usually
confused together as varieties of Aspidium Fragile
(Cystopteris) ; but most of which, in *his* opinion, are
quite distinct species ; an opinion, wherein botanists
of the Hookerian school will hardly agree with him.

We have splendid weather here, brilliant sunshine,
with a sharp east wind, which however, to-day is
much less keen than the last two days.   Thanks to
the remarkable dryness of the season, vegetation is
not much more forward than it ought to be at this
time, and not in special danger of suffering from
late frosts.   Though this place is not yet in its
beauty, it is very delightful to me, and I shall leave
it very unwillingly.

Fanny is gone over to Mildenhall for the day.
I have been walking with dear Cissy, who looks—
as she well may—much worn and shaken by the
trying scenes she went through in her long attendance

on her poor aunt; though she is beginning to rally.
Poor Lady Bunbury's sufferings in her last days are
very sad to hear of, and must have been terrible to
witness; even Cissy, as she owns, felt thankful
when it was over. How strange and sad it is, that
the separation of the soul from the body, preparatory
to its passing into another state of existence, should
generally be attended with such struggles and
sufferings. It is in vain to ask *why* it is so; one can
only trust and believe that all is for the best.

The dear children are flourishing; I saw Emmy
to-day taking her ride on a pony, under Wallis's
guidance, and cantering round the park with great
pride and delight. I hope your four are all well;
pray give them my love,

Ever your affectionate Brother,
Charles J. F. Bunbury.

Barton,
July 22nd, 1863.

My Dear Lyell,

I was very glad to see your hand-writing
yesterday, and much interested by your letter.
I wished much to know what was the special
geological inquiry which had led you away so
suddenly into France; and when I saw in the *Reader*
the account of Desnoyer's observations and speculations concerning the (supposed) marks of man's
action on "pre-glacial" bones, I thought that *they*
might have something to do with your trip. I must
own it struck me that the evidence was slight; and
this seems to be your opinion. As to Osmunda

1863. regalis, I believe that its rhizoma may be satisfactorily known from that of other British Ferns—and perhaps of other European Ferns,—by the peculiar arrangement of the vessels, which exhibit a horse-shoe shaped figure in the cross section.

The American Osmundas, and several exotic (tropical and southern hemisphere) Ferns, have much the same arrangement of vessels. I have never myself, seen Osmunda growing actually " on the sea coast," but this, is a sort of point on which Newman may be trusted.

I know that it grows near Yarmouth, in the low marshes about Fritton and Herringfleet, which must be very little above high-water mark ; so that you may very fairly say that it is capable of growing about estuaries. I am glad to hear that it has been found in the Norfolk forest-bed.

I remember Pendock very well, and the beautiful view of the Malvern hills from it, and the interesting geology of the neighbourhood. I am very glad you had such fine weather to enjoy it in ; it was wofully wet when we were there, and the leaves and fields deep in mud. Here, we had not had till yesterday, a drop of rain, since the day of the Flower Show, and we were becoming excessively parched up ; the lawn and park quite South African in colouring, and serious apprehensions of a scarcity of water. We got up our hay in a shorter time than had almost ever before been known ; only nine *working* days from the beginning of the cutting till it was all on the waggons ; the crop below the average in quantity, but excellent in quality. Yesterday evening, and 1863. this morning we have had some refreshing rain.

We had a very pleasant visit from the Maurices,* and I was very glad of the opportunity to improve my acquaintance with them. We mean to go to-morrow to Aldborough, to spend some days by the sea-side ; and on the 28th, we are engaged to go to Lord Stradbroke's, at Henham, where there is to be on the next day, a grand review of the Volunteers of the county. I hope the rain will have exhausted itself before then, but at present it is very welcome.

Fanny is pretty well, and as usual very busy. I hope dear Mary is well; give her my love. I look back with great pleasure to your visit here.

Ever yours, affectionately,
CHARLES J. F. BUNBURY.

Barton,
August 9th, 1863.

My Dear Katharine,

Our trip to the sea-side answered very well on the whole, and I had some nice botanizing at Aldborough. I wished you had been there with me. The chief things I found were these :—Lathyrus Maritimus, of Bentham (Pisum Maritimum, Linn.) a very handsome flower, growing on the shingle ridge, among the loose pebbles, south of the town, where it has grown from time immemorial. Frankenia lævis quite carpeting the ground in some parts of the salt marshes, looking like a trailing

---

* The Rev. Frederick Maurice and his wife, the sister of Julius and Augustus Hare.

1863. heath, spangled with delicate little flowers like
miniature pinks. This is a very local plant in
England, but is found in South Africa, as well as on
the Mediterranean coasts.

Spartina stricta,—a rare grass, not at all pretty ;
not found (I believe) anywhere *north* of Aldborough,
but found in many widely distant countries ; it is
one of those " especes disjointes," as Alphonse de
Candolle calls them, of which the distribution is a
great puzzle.

Statice Limonium, in great profusion and beauty ;
Atriplex Portulacoides ; Apium graveolens ; Senecio
Viscosus.

Of all these I will keep specimens for you. But I
failed to find Bupleurum Tennissimum, and some
other things which are said to grow there. By the
way, I may say that none of those which I did find
were altogether *new* to me ; but I was glad to find
them again in the same places where I had gathered
them six-and-twenty years ago.

I hope you have had some good botanizing in the
Clova mountains, and have enjoyed it. I know few
greater pleasures than a really good day's botanizing.
I hope all your dear children are quite well, and that
your Scotch tour has altogether proved agreeable
and satisfactory. Mr. Horner's seems to have been
eminently so, and it was very pleasant to read of the
well-deserved respect and honour with which he was
received at Edinburgh. Long may he enjoy them !

You will have read Fanny's account (which was a
very good one) of our visit to Lord Stradbroke's,
and of the Volunteer review there. I am sorry I

have not time to write more about it than to say 1863.
that it was a fine, and very interesting sight, and
that the party in the house was a remarkably
pleasant one. We have splendid weather ; the
harvest is going on gloriously, and the crops promise
to be uncommonly fine ; indeed, the farmers say it
will be the finest harvest that has been for many
years ; but our lawn, gardens and park are much
burnt up.

Fanny and I are both well. With much love to
Rosamond and her brothers, and to your husband,

I am, ever your most affectionate Brother,

CHARLES J. F. BUNBURY.

Pray give my kind remembrances to the Miss
Lyells.

Barton,
October 11th, 1863.

My Dear Katharine,

Well, our *grand* festivities, our private ball,
and our public ball are over, and I am alive and
well ; and Fanny, who was quite an invalid, has not
only gone through them all without breaking down,
but really seems to be rather the better than the
worse. She went through an immensity of labour,
but I must say that her exertions have been
rewarded by complete success. For myself, I
confess that a less numerous party is in general more
thoroughly agreeable to me. I get bewildered by
such numbers, but our party was as agreeable to me
as so numerous a one could be, all its elements were

1863. pleasant. Only the Herveys were a great loss, especially Sarah and George ; they were kept away by the death of their aunt, Lady Sophia. The guests staying in the house, were :—dear Minnie and Sarah Napier ; Mr. and Mrs. Berners, whom we visited a month ago, at their beautiful place, Woolverstone, near Ipswich ; Sir Charles and Lady Rowley, with two of their daughters, uncommonly nice girls ; Mr. and Mrs. Piers Freeman (Frederick Freeman's brother and sister-in-law), with their daughter ; Mr. Charles Berners, an interesting young man ; Colonel Staveley, a friend of Henry's, who has seen service in India and China ; a young Mr. Anstruther, a first-cousin of that young Anstruther, whose death at the Alma is beautifully told by Kinglake ; and lastly, Mr. Eyre, a young guardsman, a friend of Minnie's. Then we had besides, on each of the three days, several of our neighbours to dinner. The dance on Thursday was capital ; the carpet was not taken up, but covered with something (I do not know the technical name) which made it very good to dance on. The dancing was kept up with immense spirit, and the young people seemed very happy. I cannot easily imagine anything prettier than the appearance of the young ladies, especially Sarah and the two Miss Rowleys. Pray tell Susan, as she met the Miss Rowleys while she was with us, that the two who were here were *Emma and Charlotte*, and very charming girls, both of them. Mrs. Abraham was looking extremely pretty, and Mrs. Maitland Wilson, very sweet and engaging. Only imagine, I danced two dances ! ! !

But not only I, but the Vice-Chancellor,* Mrs. M. 1863. Wilson's father, who is upwards of seventy. I believe every one, without exception, joined in the dancing; it was kept up till two o'clock. Friday morning was wet, but the young ladies and gentlemen got on very merrily with billiards. We went a party of fifteen or sixteen to the ball at Bury; this I did not enjoy nearly so much as our own dance. Sir Charles Rowley and I came back at one o'clock, but the rest of the party, Fanny included, stayed till four, I believe; and the young people unanimously, pronounced it a very pleasant ball. Altogether, we were fortunate in all the members of our party; the girls pretty and agreeable, the young men very gentlemanlike; and all very willing to be pleased.

So I have given you a long account of our gaieties, You, who live in the tranquility and seclusion of London, will hardly understand or believe such dissipation!

Now, we have none with us but dear Minnie and Sarah, who are always delightful; and George Napier, who is come back from Canada, as good humoured as ever. I have not time to write any more just now as I must dress for dinner; but pray give my love to all the members of the family in Queen's Road, Harley Street, and Montague Square.

Ever your very affectionate Brother,
C. J. F. BUNBURY.

---

* Sir Richard Kindersley.

Barton,
October 15th. 1863.

My Dear Edward,

I was much interested by your account of Transylvania in your letter to me of the 16th September, and since then, I have seen your letter to Henry, from Belgrade, by which I find that you were intending to penetrate into still less known countries. I daresay Servia is a civilized country enough in its way; but I must confess, for myself, that I have much less distinct ideas concerning it, than I have concerning most parts of South America, or the South Sea Islands. I shall therefore be very glad indeed, to read your observations upon it. I must say there is a great pleasure in getting away now and then, from our excessive civilization, from railways and hedge-rows, and turnips and turnpikes, and countries of which every inch is private property; and, as I cannot in the body visit wild countries, I like to do so in the spirit. Here, since I last wrote to you, we have led in one sense quiet, and in another,—busy lives; for we have not stirred from home since our return from Norwich; but have had our house almost constantly full, and we have had some very pleasant visitors. The first party consisted of William Napier, Colonel and Mrs. Anstruther, Sir John Boileau, and two of his daughters; this was while Mr. Horner and Susan were still with us; but Leonora and Joanna away at Felixstowe. Sarah Hervey came to stay some days with us, and as she wanted to paint some large illustrations for a lecture which her

father was going to give at Bury, Susan undertook 1863.
to instruct and help her. The " bath room" was
turned into a painting room, and these two ladies
painted away for several days with such zeal and
perseverance that they would hardly allow themselves time to take a breath of air. The consequence was, that a number of capital illustrations
for the lecture were finished in an incredibly short
time.

Fanny and I both took a great fancy to the two
Miss Boileaus (Mary and Theresa), and Sir John
was very agreeable. This party broke up on the
26th of September, Mr. Horner, Susan, and Joanna
left us on the 30th, and Leonora, three days afterwards. Hardly were they gone before Fanny was
laid up with a severe cold and neuralgia, and she
was scarcely up again before the most numerous
party we have ever had, began to assemble.
Wednesday, Thursday, and Friday, of last week,
we each day sat down above twenty to dinner. On
Thursday, we had a dance in the dining room,
which was very pleasant, and was kept up with
great animation till two in the morning ; and on
Friday, we went a party of sixteen to the Bury
ball.

Altogether, our party turned out very pleasant.
I like the Rowleys very much, and the Berners too.
Now we have Minnie and Sarah with us, and
George, who is home on leave from Canada, and
the Piers Freemans. The earthquake of which you
will see accounts in the newspapers, and which
shook and alarmed all the Western and Midland

1863. counties, does not seem to have been felt at all in
this part of England.

The weather is excessively damp; very unhealthy.

Ever your affectionate Brother,

CHARLES J. F. BUNBURY.

Barton,
November 5th, 1863.

My Dear Lyell,

The book—your book—arrived quite safe,
and I am very glad that my marginal notes were of
some use to you.

I am particularly glad to hear of the expedition
of Naturalists to Palestine; it seems strange, considering what numbers of English travellers visit
that country, that so little should be known of its
Natural History, and (at least if the Quarterly
Reviewer be correct) that there should not be in the
British Museum, a single specimen of an animal of
any class, from the Holy Land.   It seems strange,
but no doubt it is easily explained, as all travellers
who go to that country have their minds and
thoughts so full of other subjects.   But I hope Mr.
Tristram and his party will dissipate our darkness in
respect of Natural History.

I thought that article (in last July's *Quarterly*) on
the Natural History of the Bible, rather well done.

About a week ago, Scott brought me some leaves
of a Fern which he had found growing on the brick-
work lining of an old well, very near here; it proved
to be the Cystopteris fragilis, a Fern which I had

never before seen in Suffolk ; and which, I believe, is generally rare in the plains of England. No doubt its spores, carried along by the winds from Derbyshire or Wales, had found a suitable station in the damp, dark interior of this old well, and had germinated, while thousands of like spores may have fallen on unfavourable ground, and perished. It is a case, less remarkable certainly, but analogous to the occurrence of Lycopodium cernuum, by the hot-springs in the Azores, and of Pteris longifolia in Ischia.

We have nearly settled to go up to town next Monday week, though I leave home with great unwillingness.

Much love to Mary,

Ever yours, very affectionately,

CHARLES J. F. BUNBURY.

BOOKS, &c., READ IN 1863.

1863.   1.—Stanley. "Sinai and Palestine."—
A very excellent and satisfactory book. I have
not read any other work which gave me such clear
and definite ideas respecting the Holy land.   There
is a clearness, a precision, a fulness of knowledge,
and a comprehensiveness of view, in all that he says
concerning it, which are very satisfactory.   He is
neither dull and prosaic, nor enthusiastic and
extravagant.   Beginning with a general survey of
the physical Geography and striking characteristics
of the country ; noticing its small extent, its remarkable position, relatively to the other celebrated
states of antiquity, its style of scenery, and its most
conspicuous productions ; he then proceeds to treat
in more detail, of the several districts from south to
north, and of the events of which they were the
scene.   He thus illustrates the Bible history, in a
very impressive and interesting manner, by showing
each personage and event mentioned in it, in their
relation to the characters of the localities.   He
remarks indeed, that the great facts and teachings
of the Gospel, were little connected with any local
peculiarities, being intended for universal application ; and he contrasts the Christian religion, in this
respect, with those of Greece and Rome, which bore
so strongly the impress of the particular countries in
which they originated.   Yet, at the same time, he

shows in a striking manner, how some parts of our 1863.
Lord's teaching, and principally the parables, are
illustrated by the objects which are still daily before
the eyes of the inhabitants of Galilee and Judea.

Dr. Stanley is not by any means enthusiastic (as
some travellers are) about the beauty of Palestine
generally. He represents it,— at least the hill
country of Judea—as on the whole, a rather
melancholy and monotonous style of country, though
certainly appearing to advantage, by contrast with
the absolute desert on the south and east of it.
A tract of arid, bare, grey limestone hills, rather
tame and monotonous in outline, with few conspicuous points or marked features ; such is the
impression he gives us of the aspect of Judea in
general.

This is relieved by, and contrasted with
the verdure and fertility of the plain of Esdraelon,
the vales of Samaria and Gennesareth, and a few
other spots where abundant springs nourish a fresh
and luxuriant vegetation. The dense thickets along
the banks of the Jordan, and the forests on the sides
of Carmel and Hermon, must also be very striking
to an eye accustomed to the general aridity of
Palestine.

Having finished his survey of the country with
Lebanon and Damascus, Stanley notices briefly,
the several so-called "Holy places," and the
superstitions and fables which the monks have
attached to them. He concludes with some excellent observations, showing how the real, deep,
permanent religious interest with which Christians

1863. must always regard Palestine, is independent of all
these local superstitions.

2.—Falconer. " On a New American Fossil
Elephant." *(Nat. History Review*, No. 9.)
A most elaborate, careful, and important paper,
full of information. Professing merely to be an
account of a species of Fossil Elephant (Elephas
Columbi, Falc.) newly discriminated from the Mammoth, it is really a most copious and careful survey
of the natural history of Elephants, recent and
fossil. What I notice in it more particularly, are the
discussions on the geographical range of the new
species and of the Mammoth,—of the geological
position and range in time of the Mammoth ; and of
the question, what were probably the earliest head-
quarters of that species ? the very full investigation
of the question, whether there be more than one
existing species of Indian Elephant, in which he
shows that the Elephas Sumatramus is not a distinct
species from the E. Indicus ; and the curious
account of the food of the Indian Elephant in a
wild, and in a tame state, with the striking observations on the effects produced by the change of diet
on the teeth.

There is much worth noting and remembering in
this paper.

3.—Thiers.   " Hist. du Cons. et de l'Empire."—
Books 28, 29, 30.—

These three books, which make the eighth volume, 1863. are almost entirely occupied with the affairs of Spain, which are related at great length, and with rather tedious minuteness of detail, from the beginnings of Napoleon's designs against that country, to the consummation of the treason at Bayonne, and the entry of Joseph Buonaparte into Spain. Thiers dare not attempt openly or avowedly to vindicate Napoleon's conduct towards Spain ; but he seeks indirectly to excuse or palliate it, by dwelling and expatiating at great length on the abominations of the Spanish court and Government. And, whatever excuse may be furnished to the ambition of a conqueror by the vices, follies, and cowardice of a court and a Government, such excuse, we must admit, was afforded to the fullest extent, by the rulers of Spain to Napoleon. The temptation was no doubt great ; Napoleon, looking only to the court and Government of the country, overlooked the people entirely, and this error brought on him the retribution which he had deserved.

4.—Charles Lyell. "Geological Evidences of the Antiquity of Man."—

This is a great and admirable work. To my thinking, by far the best thing he has written since the original *Principles of Geology*.* And there is a vigour and animation, a freshness and clearness, which remind me of the *Principles*, and recall to mind the delight with which I first read that classical

*Read in 1831.

1863. book. This also, though perhaps in a somewhat lesser degree, may be ranked with the former, among the works which constitute permanent landmarks in the history of a science.

Lyell begins by giving a very clear summary of all the evidence relating to the ancient races of Man, that has been afforded by the remains of the " Lake dwellings " of Switzerland, and by the "Refuse heaps " and peat bogs of Denmark.

He next describes the human remains· found in caves, in situations showing their great antiquity, and often in company with the bones of extinct mammals in various localities in France and Germany ; some of these causes were known to geologists at least five-and-twenty years ago, but were looked upon with much suspicion (see the 5th edition of his "Principles"). He enters into full anatomical details (on the authority of Huxley) respecting some of these human skulls found in caverns in Germany. Then he proceeds to treat of the celebrated " flint knives " and " flint hatchets," found in the gravels of the valley of the Somme, and of some parts of England ; and dwells upon them at considerable length, showing the evidences of antiquities afforded by the geological position in which they occur. He hardly seems to think it necessary to prove that these implements are of human workmanship ; and indeed I suppose there are now not many who would dispute this point.

So much then appears to be proved beyond a doubt—that the human race co-existed in these parts of the earth with numerous large mammifers, which are not only now extinct, but appear to have 1863. become so before the strictly historical period; as well as with others which still exist. And from the levels, the drainage, and other points in the physical geography of the districts, it is argued that we must carry back the dates of these remains to an antiquity far exceeding what has been usually assigned to man.

So far, Lyell deals strictly with his professed subject, "the Antiquity of Man;" next he proceeds to show, by an exposition of all that is known concerning the Glacial Period, how vast a length of time must be allowed to that period; and consequently, how enormously great must be the antiquity (according to our ordinary measures of time) of those numerous species of animals and plants which existed before the Age of Ice, and are still existing. These chapters on the Glacial phænomena, though extremely well done, are much more difficult than the earlier ones. They are written with so much conciseness that they require very close attention, involving as they do, a highly complicated and perplexing series of facts. There are few things in geology, I think, more astonishing than the succession of enormous movements and changes, of which the history fairly deduced by scientific reasoning from the observed facts, is expounded by Lyell in these chapters. It strains one's imagination to conceive the length of time that must have been required for so many and so great changes of level, upward and downward, (all probably taking place very gradually), and such

1863. changes of climate as are shown to have taken place during the course of the Glacial Age. And yet, not only are very many of the now-existing species of Mollusca found in the beds which must have been deposited *before* the Glacial Age, but there seems to be clear evidence that the Scotch Fir also, and several of our common North European plants, were of *pre-glacial* antiquity. The account of the "forest bed" in the Norfolk cliffs is especially curious in this respect.

The third part of Lyell's book, including the last five chapters, is devoted to an examination of the great question of Species, and of the Lamarckian and Darwinian theory of variation. This is a subject on which, as had long been well-known to his friends, Lyell had entirely changed his opinion since the days when he treated of it in his Principles; and therefore he felt himself bound to go fully into it, although it is perhaps not very directly connected with the primary subject of this present work. He has now, most frankly and fairly avowed his change of opinion, and gives, in chapter xxi. of this book, a clear summary of Darwin's theory, and of the main arguments on which it rests. He does not indeed look on it as proved, and therefore does not express himself on it with the same positive confidence as Darwin, Huxley and some others ; but it is very clear which way he leans. The chapter in which he compares the development and genealogy of species with the development and genealogy of human languages, is remarkably ingenious and striking.

In the last chapter, Lyell discusses the great and 1863. much-vexed question of the degree of affinity of Man, with the other Mammalia, with the Apes in particular, and how far it is probable or otherwise that the human species was formed by the "law of development or variation" out of some of these inferior animals. In so far as relates to the affinities of bodily structure, and especially to the characters of the brain and of the foot, Lyell here follows Huxley implicitly, and sides altogether with him against Owen. The summary that he gives of the famous "hippocampus" controversy, is known to have excited Owen to great wrath.

The latter part of this chapter is written in an excellent spirit, full of reverence and of a truly religious spirit.

5.—Huxley. "Man's place in Nature."—

This small book, which was published immediately after Lyell's, may be considered in some degree as a sequel to it; for it treats essentially of the same subject to which the last chapter of "The Antiquity of Man" is devoted. Huxley here gives, in a very clear and lively style, an exposition of his views concerning the close affinity of Man to the Apes, and of his favourite doctrine of Variation and Development. The first part of his book, in which he gives an account of the structure and characters of the "anthropoid" Apes, and a summary of all that is known of their habits and manners, is extremely well-done. He seems also to

1863. be successful (supposing him correct as to his facts) in showing that the structural differences between Man and these "higher" Apes, (the Gorilla in particular) are less than between the Gorilla and some of the lower forms of Monkeys. He evidently believes that there is a graduated scale of intellectual and moral powers as well as of physical structure, connecting Man with the Monkeys; and that the differences are rather in degree than in kind.

6.—Kinglake. " Invasion of the Crimea ;" Vols. i. and ii.—

Whatever may be the faults that may be found with this book, it is certainly very agreeable reading, and leaves a strong impression on the memory. In fact, I have seldom read a more entertaining history. Kinglake's style—or rather, perhaps, his manner of arranging and handling his subject, is very peculiar, distinctively his own ; almost as peculiar, though not as uncouth as Carlyle's. To impartiality he can certainly lay no claim ; his dislike to the French, and his special animosity against their Emperor and his adherents, are so very glaring, and appear so continually, and in such exaggerated forms as to prevent us from feeling any confidence in his representations wherever our allies are in question. This animosity leads him also into what appears to me an error in a literary view ; it leads him to occupy much the longest chapter of the first volume—considerably above a hundred pages

—with a very copious, elaborate, and highly-wrought narrative of Louis Napoleon's *coup d'etat* in December, 1851 ; and of the massacre on the boulevards of Paris. This is certainly striking and vigorously written, in itself, though too much in the tone and manner of a political pamphlet ; but it is not properly a part of the history, nor directly connected with it, and the scale on which it is executed seems quite out of proportion to the rest. With the exception of this enormous episode, the first volume is filled with the negotiations preliminary, and leading to the war. This is all taken from the official papers, and it is remarkable how interesting, and even entertaining, Kinglake has contrived to make the substance of sundry *blue* books, which he has here brought before us in a most readable form. The chapter in which he gives an account of the dispute about the " Holy Places," is a very amusing specimen of his way of treating these subjects. The fairness of the representation he has given of these diplomatic combats has been strongly disputed, and it must be so far admitted, that his evident and strong bias against the French Government may readily be supposed to warp more or less his views of the whole subject. But his account of the helpless sort of way in which the English Government " drifted " into the war, and of the manner in which the Czar and Europe generally might have been misled by the declamations of the peace party into totally false notions of the general state of thought and sentiment in England—appears to me, very good and very true.

1863. Some of his portraits of public men are very lively and entertaining; they are always drawn in very strong colours, whether favourable or the reverse, and often border on caricature. It is impossible to believe that the Emperor Napoleon is so poor a creature as Kinglake represents him; and yet when one compares the part he played at Strasburg and at Boulogne with his subsequent career, it must be owned that he is one of the most enigmatical characters in history. It is when we come to the actual invasion of the Crimea, in the second volume, that Kinglake appears to the greatest advantage. The whole account of the landing, of the march to the Alma, and of the battle, is a fine piece of narrative. The story of the battle is told in a very remarkable and striking manner; unlike any other narrative of battle that I remember to have read. It is immensely long, about 300 pages—and as may be supposed, very copious and almost minute in detail; strikingly picturesque and vivid, and perhaps making all the more impression from the copiousness of detail and explanation on many points which are unfamiliar to a non-military reader, but which a professional military writer would not think of explaining.

Some of the incidents are told with much simple pathos; such as the death of young Anstruther and that of young Eddington by the side of Major Champion.

Perhaps it may be said that the profusion of detail rather interferes with a clear conception of the general idea of the battle. Certainly the

impression we derive from the whole, is that the 1863.
battle was very unskilfully fought on the part of the
Allies—and fortunately still more so on the part of
the Russians.

Lord Raglan, though Kinglake's favourite hero,
is clearly shown to have played the part of anything
but a good General.

———

7.—Bates on "The Insect Fauna of the Amazon
Valley." Linnean Transactions. Vol. xxiii.

A curious paper on a beautiful tribe of butterflies,
very characteristic of Tropical America, the Heliconidæ—Heliconii of Linnæus. The remarkable
point, and the chief subject of the paper is that as
Mr. Bates seems to have been the first to observe
certain other butterflies, inhabiting the same region,
but belonging by their essential characters to quite
different tribes, bear in their colouring and general
appearance so striking a likeness to certain abundant
species of these Heliconidæ as, when seen on the
wing, to be almost undistinguishable from them.

These Mr. Bates calls *mimetic* resemblances, and
he illustrates and follows them out in detail in a
careful and interesting manner. As the Heliconidæ
(the insect imitated) are very abundant in their
special districts and seem to be in a very flourishing
condition, and the particular species (of the Pierida
tribe) which imitate them, are comparatively very
rare, he concludes that the design and purpose of
this curious imitative variation is to preserve the
weaker species by an assumption of the dress as it
were, of the more powerful one.

1863.   8.—J. D. Hooker. " On the Genus Welwitschia."
Linnean Transactions.   Vol. xxiv.—
A very remarkable paper: a most careful and
elaborate description, beautifully illustrated of the
whole structure and characters, external and
anatomical, of that marvellously strange plant from
South-Western Africa, the Welwitschia.   It is one
of the most masterly even of Joseph Hooker's
works.   The plant itself (of which since reading
this paper I have seen the specimens at Kew) may
safely be said to be the most uncouth and shapeless,
and (in general appearance at least) the least
beautiful of all flowering plants as yet discovered.
Its affinity is here clearly shown to be with the
Gnetaceæ ; a small family, of which the only two
genera previously known, Gnetum and Ephreda, are
themselves so dissimilar, and in many respects so
peculiar and exceptional, that one is led to view
them as the last remnants of a family that is
disappearing.

There is a good analysis of this paper in the
Natural History Review, number 10, (April, '63), in
which the especial peculiarities and anomalies of
Welwitschia are briefly and clearly pointed out.

9.—Stanley. " Lectures on the Jewish Church."—
This is a singularly interesting and delightful
book ; really a fascinating book.   I never read any
other work—not even Milmans " History of the
Jews,"—which made me feel as this does, the
reality of the Old Testament history ; which brought
that history clearly and vividly before my mind, and

made it possible for me to take a real lively interest 1863.
in it. Dr. Stanley, with his fine poetical imagination,
his thorough knowledge of the country, and of the
present as well as past manners and customs of the
people inhabiting it, united to ample learning,
has overcome the associations of something (I must
say it) like disgust, with which the Puritans and
Methodists, and the dull droning of " lessons " in
Church, had (to me at least) enveloped the names
and events of Jewish history. I know that some
people of both the more extreme parties in our
Church, accuse Stanley of "disbelieving Scripture,"
and nourishing doubt in the minds of his readers.
I can safely say for myself, on the contrary, that he
has made the Hebrew history, not less, but much
more credible to me. He has brought it out of the
region of formulas and shadows into that of realities.
Stanley is evidently not one of those who make
an idol of the Bible ; not one of those who require
us to believe in the literal truth of every statement and every word contained in it. But instead
of occupying himself in unsatisfactory verbal or
numerical criticism, he—very wisely, I think—
devotes his great powers to the development,
explanation and illustration of the historical truths
and moral lessons in which these records are so
rich. He touches merely incidentally, and by
passing allusions on the many difficulties as to
numbers, dates, and other matters which are to be
found in the Hebrew Scriptures, but he brings
vividly before us the patriarchs, heroes and sages of
the history, shows us the individual differences of

1863. their characters, the scenes and circumstances in
which they acted, the people with whom and against
whom they acted, makes us feel with and for
them.

10.—Bates. "The Naturalist on the River
Amazons."—

This treats of the same region as Wallace's work
which I read last year, and is a still better book
than that.   Indeed it is by far the best book of
natural history travels that I have seen since
Hooker's Himalayan Journals.   In one respect it
has a decided advantage over everything that (so
far as I know) has hitherto been published on the
natural history of tropical countries : unless indeed
it be Sir Emerson Tennent's "Ceylon," which, on
the other hand, is inferior in scientific depth.   Mr.
Bates lived not less than eleven years together in
the region of the Amazons, and most part of the
time in the wild interior, in spots which had before
been merely visited in passing by scientific
travellers.   For four years and-a-half his head
quarters were at the remote village of Ega, near the
junction of the river Teffe with the Upper
Amazons, 1900 miles from the mouth of the great
river, a spot which, I believe, hardly any men of
science, except Von Martius and his companion had
ever before visited.   The consequence of his
allowing such ample time for the exploration of
these interesting countries has been, that besides
discovering a vast number of new species, he has
been able to give us far more interesting and

satisfactory accounts of the habits and manners of
many animals than can be found in any other book.
His biographies of the Saübas or leaf-carrying
Ants, of the several species of "foraging" Ants
(Eciton) of the Termites, and of many kinds of
Monkeys, are worthy of Gilbert White. The work
is rich also in important observations on the
geographical distribution and local variations of
animals. Some of the facts recorded under this last
head are very curious :—as to the manner in which
certain supposed species—of butterflies for instance,
are confined to a certain limited tract of country,
between particular rivers, and in another tract, not
far distant, nor under very different local conditions,
are replaced and represented by other forms,
which having definite and apparently permanent,
though not very important differences in colours and
markings, are generally admitted as distinct species.
Mr. Bates is a zealous Darwinian, and endeavours
to explain these facts according to that theory. I
cannot say, for myself, that his explanations are
altogether satisfactory to me, as to the precise mode
in which the changes have been effected : but I
cannot help believing with him, that in very many
cases, what are called allied, are representative
species, really sprung from one common stock,
modified by the action of natural or secondary
causes.

Mr. Bates points out that the fauna of the lower
Amazon valley (that is, from the mouth of the river
up to the junction of the Rio Negro), has no close
relationship with that of Southern or Middle Brazil,

1863. but is closely connected with that of Cayenne and
Demerara, or the coast region of Guiana. He says
that Von Martius, when he visited the Lower
Amazons, coming from the South was much struck
with the dissimilarity of its animal and vegetable
productions to those of the rest of Brazil. He also
remarks that the comparatively high land which
borders this valley on the South, and which extends
down the Tapajos river nearly to its junction with
the Amazons in the neighbourhood of Soutarem, is
materially different in soil, climate, vegetation and
animals, from the flat, low country which generally
borders the great river ; and that it appears to be a
continuation of the table land of Central Brazil. It
would be interesting to know whether (as is probable) these observations hold good with regard to
the plants as the animals of those regions. This
might be ascertained by a comparison of poor
Spruce's collections on the Amazons with those
of Von Martius and Gardner in the interior of
Brazil. It is certain indeed, that some conspicuous
plants (such as the Cecropia) are common to the
West Indian Islands and Rio de Janeiro, and that
there are some which range from Guiana even to
South Brazil ; but it is probable that in the main,
Guiana and Amazonia would be found to be distinguishable as a separate botanical province from
Brazil proper.

Mr. Bates is a more agreeable writer than Mr.
Wallace, he writes with more animation and zest,
and seems more strongly impressed by the charms
of tropical scenery and productions, of which he

gives us many spirited and life-like pictures. He 1863. writes as one who had thoroughly entered into the enjoyment of the scenes and pursuits amidst which he passed so many years; indeed he acknowledges that it was with considerable reluctance and misgiving that he quitted those countries to return to Europe.

11.—Stanley. " Lectures on the Eastern Church."—

Less interesting and attractive than the "Lectures on the Jewish Church," because the subject is so much less interesting; but still more agreeable reading than I should have thought almost possible with such a subject. The two chapters on Constantine and Athanasius are very striking. Indeed the whole story of the Council of Nicæa is told in a remarkably clear and interesting manner ; the opening scene is brought vividly before us, and the characters and appearance of the chief personages are represented with great spirit. I have listened with more interest to the reading of the Nicene Creed in Church, since reading in this book of Stanley's, the history of the debates and discussions of which it was the result. But I must say that the early Church even so soon after its release from heathen persecution, does not appear to advantage in the Council of Nicæa.

The bitterness of spirit, the want of Christian Charity towards opponents, the fierce debates turning on the most inconceivably minute shades of difference in expression : the animosity shown in

1863 discussing questions which could have no possible
bearing on practice, and of which no mere mortal
could really *know* anything ; these show a sad falling
off from the spirit of the Apostolic times.   Yet
there were great men among those Nicene Fathers.
The chapters on the Russian Church, though of
course much less interesting, are likewise curious.
It is curious to see the operation of Christianity on
a barbarous people and their barbarous rulers.
And amidst all the ignorance and superstition and
fanaticism which disfigure Russian Christianity, it
is interesting to see that the only moral checks
which the tyranny of the Czars has ever endured,
have proceeded from Christian preachers.

---

12.   Gibbon.   "Decline and Fall," &c.   Chapter
     xvii.—

After reading Stanley's "Eastern Church," I
thought it would be interesting to read those
chapters of Gibbon which relate to the time of the
Nicene Council.   I read accordingly the reign of
Constantine and the two chapters (xx. and xxi.)
treating of the establishment of Christianity as the
State religion, and of the Arian and other sects ;
and then was led on by the interest of the subject
and by Gibbon's powers of narrative to read the
whole of the reign of Julian, and thence onward to
the end of that of Theodosius.

The character of Constantine appears to much
less advantage in the pages of Gibbon than in those
of Stanley ; and this not merely because of the

different light in which the two authors may view his conversion and its results; but rather because Stanley's object is chiefly to show the Emperor in his relations to the Church rather than to the State.

Constantine seems to have contributed more than any other Emperor to the ruin of the Empire; not by his conversion, nor by his removal of the seat of the Empire; but by completing and systematising the establishment of despotism. Under the earlier Emperors the Government was an irregular and incomplete despotism, retaining most of the forms and something even of the spirit of the Republic; liable to occasional extravagances of tyranny, but on the whole a good deal checked by old forms and maxims and traditions. Diocletian seems to have begun the work of converting this into a regular and thorough despotism on the Oriental pattern; Constantine completed it. He destroyed the last vestiges of freedom, organized and established a complete system of autocratic tyranny; at the same time corrupted and ruined the discipline of the army by systematic favouritism.

The fine, but unequal and ill-balanced character of Julian is very well displayed by Gibbon. He excites more of our sympathy than of reasonable and deliberate approbation. He is the only one of the later Emperors, I think, in whom one takes any personal interest.

The story of Julian's fatal march against the Persians, is a fine piece of narrative. The general description of the manners of the Tartars, the

1863. passage of the Danube by the Goths, and the destruction by them of the Roman Army in the battle of Hadrianople, are very masterly.

13.—On the Lignite Formation of Bovey Tracey, Devon, by W. Pengelly and O. Heer.—

An important paper, reprinted from the Philosophical Transactions, with a historical introduction by Pengelly, giving an account of the observations and opinions of previous writers on the Bovey coal formation. In the body of the paper, Pengelly describes clearly the extent, position, physical characters and stratification of the deposit, and points out the circumstances under which it was probably formed ; while Heer details with minute accuracy the character of the vegetable remains found in it. Though of small extent, the Bovey deposit is of much geological interest, since with the exception of two other still more local or fragmentary formations, it is the only representative in Britain of that important series of beds which are generally called Lower Miocene, but by some geologists Upper Eocene, and which include most of the brown coal of Germany. The other two examples are the Hempstead beds in the Isle of Wight, and the leaf bed discovered by the Duke of Argyll in the Isle of Mull. The Bovey beds occupy a small plain, or lake-like expansion at the foot of Dartmoor, on its eastern side ; the extent of the plain being about six miles by four. They are surrounded by rugged and elevated districts of palæazoic rocks and granite: and they appear to 1863. have been formed in a deep and tranquil fresh-water lake. The prevailing plant is a Conifer Sequoia Couttsiæ, which is very completely de-scribed by Heer, the evidence as to its nature and characters being more satisfactory than one can often obtain in the case of fossil plants. It was so nearly related to the existing Sequoia gigantea (Lindley's Wellingtonia) that the differences seem but slight.

This memoir is illustrated by very beautiful figures of the plants.

14.—Jamieson. On " The Parallel Roads of Glen Roy;" in *Quarterly Journal of the Geological Society*, vol. xix., part 3.—

A very clear account of the phenomena of that curious and celebrated locality; much fuller and therefore clearer than that given in Lyell's works. Mr. Jamieson's principal conclusions, which seem to be supported by strong reasons, are these:—

That the parallel roads were the beaches of fresh-water lakes.

That these lakes appear to have been caused by glaciers damming up the mouths of the valleys and reversing their drainage.

That the date of these lakes is posterior to the great " land glaciation of Scotland."

15.—Milman's "History of Latin Christianity."—

This work is professedly a sequel to the author's " History of Christianity," and this first volume

1863. brings down the history of the Western or Latin
Church, from the time when Christianity became,
under Constantine, the religion of the State, to the
complete ascendancy of the Monkish system, when
at the end of the sixth century "Monasticism
ascended the papal throne in the person of Gregory
the Great."

Naturally, in order to illustrate the course of the
Latin Church, and to show the points of difference
between the two, the historian finds it advisable to
give, incidentally, outlines of much of the history of
the Greek Church; and he begins with a brief
notice of the position of the Roman Christians,
relatively to those of other parts of the Empire,
before the triumphs of Christianity.

The period of history embraced within this
volume is a curious and important, but not in any
point of view an agreeable one.   Whether we look
to the rapid degradation of Christianity under the
influence of the Byzantine Court and of the eastern
monks; to the fanatical sectarian fury excited by
unintelligible differences about incomprehensible
dogmas; to the Councils in which authoratative
decisions on questions beyond the reach of human
reason were obtained by violence and intrigue; or
to the miserable condition of the Roman world,
suffering every calamity that internal decay and the
ravages of barbarous invaders can inflict; or the
ferocity of their invaders; in any way it is not a
cheerful and agreeable spectacle. Yet we cannot
help seeing that Christianity, soiled and perverted
though it was, was the one redeeming and civilizing

element in that chaos; and that even the Monkery, which was the form that Christianity then generally took, however different from primitive or from enlightened Christianity, was yet well adapted to the wants of the dreadful times. Milman points out clearly the operation of the Christian religion on the conquering Teutonic nations; the causes which made its progress among them, easy and rapid; and the manner in which it became modified by the natures on which it acted.

He shows very well how, amidst the chaotic ruin of the Roman Empire, the Clergy alone "stood between the two hostile races in the new constitution of society: the reconcilers, the pacifiers, the harmonizers of the hostile elements."

The nature of that strange anomalous form of human society, the monastic system as it grew up amidst the wrecks of the old society and the struggling confusion of the new, is well illustrated, especially by a pretty full sketch, which Milman gives of the career of the great S. Benedict. A most curious specimen it is. One can hardly imagine anything more alien from all our habits of thought and feeling, from all our habitual notions of human life and human nature.

Nature indeed seems to be the last thing thought of; such men as the biographers of Benedict seem to have viewed the laws of nature as existing only to be set aside. A perpetual succession of miracles is considered as the normal state of things in the life of a Saint.

1863.   16.—Geikie, on the Ancient Glaciers of Scotland.
—*(North British Review.* November, 1863).—
A very clear and well-written exposition of the
glacial theory in its present form, as far as relates to
Scotland; that is, of the series of changes which,
as it appears, can be proved to have taken place on
the surface of (what is now) Scotland, during the
course of the age of Ice. A very wonderful history
it is. I must assume that the author's facts (which
I have had no opportunity of verifying) are correct.

17.—Froude's "History of England." Vol. vii.—
This volume of 540 pages carries us through only
five years of the reign of Elizabeth, from November,
1558 to December, 1563. We might be inclined to
think that a whole volume of 540 pages is too much
to allow to so short a space of time, but the political
affairs, the position, and interests of England,
relatively to other nations during that period, were
so marvellously intricate and perplexing, that probably they could not have been explained in such
a manner, as to be either interesting or intelligible
in a less space. At least, there would seem to have
been no middle course, between attempting a mere
grand outline after the manner of Hume, and
entering into copious details, as Froude has done.
The intricate complexity of the foreign relations
of England during the earlier years of Elizabeth,
and the tangled web of intrigues in which she and
her statesmen were plunged, are indeed quite ex-
extraordinary. The peculiar relations in which she

stood to the King of France and to the different 1863. factions in that country;—to the King of Spain, who longed to crush her as a heretic, yet found himself obliged in sound policy to support her, lest France should be strengthened ;—to the Queen of Scots, her dangerous rival and claimant of her throne ; — to the Reformers in Scotland, whom policy obliged her to support against the French power, while she felt no sympathy with their religious views, and looked on them with dislike as rebels ; — these were sufficient to make the situation a very extraordinary one. And when to these are added her coquetries and uncertainties or dissimulations about her marriage, and her frequent hesitation (at least apparent) between the Romanist and Protestant faiths, the complication becomes something quite bewildering.

As far as this volume goes, I must say that it gives me a very unfavourable opinion of Elizabeth's character, worse than ever I had before. Her position was indeed a very difficult one, but her vacillation and her duplicity were outrageous.

Often one feels doubtful whether there was more of real irresolution or of insincerity and sheer false- hood in her conduct ; but both seem to have had their share ; and one feels it difficult to allow that one who showed both these qualities in so great a degree, could deserve to be thought good or great. The newest and most remarkable part of this history consists in the large extracts given by the author from the dispatches of the Spanish Ambassa- dors in England, to which he obtained access at

1863. Simancas. Very curious and very interesting these
extracts are, especially those from dispatches of
Alvarez de Quadra, Bishop of Aquila, evidently a
man of very remarkable ability, a most accomplished
diplomatist and sagacious observer. All his remarks
on the characters of statesmen and the position of
parties in England, as well as those on Elizabeth
herself, are very instructive. Perhaps while he
thoroughly understood and calculated on the weaknesses and faults of her character, he did not
sufficiently perceive the real strength which lay at
the bottom of it. He certainly gained a considerable though varying degree of influence over
her. It is very curious to see how she is swayed
backwards and forwards as the influence of the
Quadra and that of Cecil by turns gained the
ascendancy ; like the evil and good angel in
some of the mediæval legends. Certainly, I think
we may say that under Providence, it was mainly
owing to William Cecil that Protestantism became
firmly established as the religion of England.
Elizabeth's aversion to the more decided forms
of Protestantism was very apparent ; her attachment to Protestantism at all, except as a mere
denial to the Pope's supremacy, appears very
questionable.

Perhaps her pride and spirit would in any case
have kept her from submitting to the authority of
the Pope, but it appears probable that, had it not
been for the influence of her good genius Cecil,
she would have been glad to establish a modified
Romanism as the religion of the State. The danger

was the greater because Leicester, by whom she 1863. was so strangely fascinated, at this time threw his weight into the scale inclining towards Spain and Rome. Later in his life, if I am not mistaken, he favoured the more Puritanical party, but in these years his weak and selfish character appears to have fallen under Romanizing influences. With regard to Elizabeth's perplexities and duplicity about her marriage, I am inclined to believe that the truth is this:—not that she had any absolute and settled aversion to marriage in itself; but that the man of her heart—the only man she would willingly have married was Leicester—that the advice of her best counsellors, her knowledge of the general feeling against him prevented her from marrying him; and that she could not reconcile herself to marrying any-one else. The difficulties of her position, and the pressure of the popular feeling, repeatedly made her *almost* determine to accept for her husband some foreign prince; but her personal feeling always prevented the fulfilment of the sacrifice.

The mode of composition or arrangement of the matter of this book is rather peculiar. It might be easy to find fault with the plan that Froude has adopted, of introducing numerous and long extracts from the original documents of the time, and even whole dispatches into his text, instead of digesting and incorporating the substance of them with his narrative, and throwing the originals into the form of notes.

This latter method would doubtless be more in accordance with the practice of the most

1863. classical writers ; but Froude's mode of telling
the story seems to me to give a more lively
impression of reality ; to bring us more into
personal acquaintance and contact with the actual
men and women of those times.

LETTERS.

Barton,

January 3rd, 1864.

My Dear Katharine,

1864.   I had meant to write to you on the last day
of last month, to wish you and yours a happy new
year, but something or other prevented me : however, I hope it is not now too late to express my
hearty and earnest good wishes for you, that you,
and all who are dear to you, and your children in
particular, may enjoy health and happiness
throughout the year that has now begun. For
myself, I have every reason to feel devoutly
thankful for the situation in which dear Fanny and
I find ourselves, and for the blessings with which
I am surrounded. I am afraid the year is likely to
be one of struggle and sorrow to very many, for
the prospect of peace, on the Continent generally,
appears, (as far as I can judge) very slight indeed.
Winter seems to have come at last with the new
year, for though no actual snow has yet fallen,
yesterday was really like a winter's day, and to-day
still more so, and it looks like the setting in of a
real hard frost. Here is a list of the plants which
were in flower in the open air in our garden on the

first day of the year.   Jasminum nudiflorum, 1864.
Eranthis hyemalis (the so-called Yellow Aconite),
Daphne Laureola, Viola odorata (in profusion),
Chinese Chrysanthemum, Anemone coronaria,
Primrose, Cynoglossum omphalodes, Wallflower,
Laurustinus, Arbutus, China Rose.   At Mr.
Rickards' a few days ago, we saw the Coronilla in
profuse blossom out of doors, against the wall of
the house.

Have you read Froude's two new volumes, the
7th and 8th ?   I have not yet finished the 8th,
which is the more striking of the two; all the
history of Mary Stuart's proceedings is most
strikingly and powerfully told in it : but both are
very interesting, and full of most curious matter
very ably put together.   I have skimmed over
"Capt. Speke's African Travels," but not found
them very instructive : there is a vast deal of a
sort of Court Journal and *Chronique scandaleuse* of
the Court of Uganda, which is perhaps not exactly
what one most wishes to know.

Pray give my love to your husband and the dear
children and all your belongings.   I hope we shall
to-morrow hear that Mr. Horner is well again.

Believe me ever,

Your most affectionate Brother,

CHARLES J. F. BUNBURY.

Barton.
January 24th, 1864.

My Dear Katharine,

I have a letter to thank you for, which gave me much pleasure, and since then I have heard from time to time of you and yours, and have been very glad to hear of the various family parties and gaieties. I hope you and your children have continued well; the fine mild weather which has followed the sharp cold seems to have disagreed with many people. I have been very sorry to hear of Mr. Horner's illness, and that he is still so weak; but I trust he will soon gain ground. I can feel for his vexation at being so long kept away from the Geological Society.

We have been quiet enough this month, as far as society is concerned, and have indulged to the full our organ of stay-at-home-ativeness; but I have been a good deal worried by troublesome business relating to Mildenhall, which has consumed time and thought, and interfered with intellectual occupations. However, I hope it is now coming to a satisfactory conclusion.

Since finishing Froude, my curiosity having been excited about Mary Stuart, I have been reading Robertson's "History of Scotland," and have now gone through a volume and a half of it. At first I did not relish the style, it even seemed insipid after Froude; but by degrees I got "into the way of it," and have gone on reading with great interest and pleasure;—though I must own the style even now appears rather stiff and formal, too uniformly worked up and polished, and there are too many 1864. trite moral reflections for my taste. The history in itself is one of the most romantic imaginable: absolutely crowded with extraordinary characters and extraordinary events. What a fierce turbulent race the Scots of that age were! Almost every one of those who took a prominent part in the public affairs of Scotland, in the latter half of the 16th century, came to a tragical end. It is strange that John Knox should have been one of the few exceptions.

I have as yet only looked through the last part of species Filicum, — not yet begun to name my specimens after it. I see that Sir William treats Phlebodium, Drynaria and all the other modern genera of that group, even Niphobolus—as sections of Polypodium.

The work altogether is one of enormous labour and difficulty, and probably no other man living could do it as well as he has; but still I must confess that, as it appears to me, it leaves much to be desired. His genera—many of them, at least, and Polypodium in particular, are purely artificial groups, and seem to be merely temporary contrivances to serve till some better principle of arrangement can be discovered. Even as to species, in spite of his immense knowledge and industry and care and skill, it often strikes one that he unites under certain species so many different plants apparently so widely distinct that one wonders on *what* principle he admits any distinct species at all. But after all, it is ten times

1864. easier to find fault than to say how it could have been done better.

The week of severe weather at the beginning of this month seems to have done no mischief here, not even killed the flower buds of the Laurustinus. The ground is now spangled with the little yellow Aconites; the Crimean Snowdrop is in blossom in the beds before the house; the common Snowdrops are peeping out, and the leaves of the Honeysuckle are beginning to appear. These last few days have been beautiful.

Fanny is much occupied in watching little birds, which she has at last succeeded in attracting to the neighbourhood of the drawing-room windows; some bits of mutton bone have proved the lure which has attracted three different sorts of Titmice; charming little birds they are, and it is very amusing to watch their manœuvres.

With much love to your husband and children,

I am ever,

Your very affectionate Brother,

CHARLES J. F. BUNBURY.

[In the end of January or the beginning of February, the Charles Bunburys went up to London to 48, Eaton Place, on account of the illness of Mr. Horner, who after lingering a month, died on the 5th of March. Soon after the funeral they returned to Barton, where the repose was very comforting after this time of sorrow].

Extract from a letter of Sir C. J. F. B. to Madame 1864.
Byrne.

16th March, 1864.

I shall always feel it a privilege to have been so long and so intimately connected with so excellent and wise a man as Mr. Horner. I shall always feel that I am, or ought to be, the wiser and the better from having known him so intimately, and the recollection of his kindness, and of the affectionate regard that he showed towards me will always be deeply gratifying.

And when the first natural impulse of grief is alleviated by time, I think his children, and you his sisters also, will feel comfort in reflections on his long honourable, useful and happy life.

I have never myself known a better man, nor one of more universal and unfailing indefatigable kindness. His active benevolence never cooled in the least—the warmth of his domestic attachments, nor did his strong affection for his family and friends check his enlightened ardour for the general good of mankind.

Extract from a letter of Mr. Poulet Scrope* to Sir Charles Lyell.

" Let me begin by condoling with you on the loss "you have sustained in common with a large circle "of friends and admirers.

"So admirable a person in mind, manners and

---

* Mr. Poulet Scrope was a distinguished geologist and brother of Mr. Poulet Thompson (Lord Sydenham), Governor of Canada in 1839 (?)

1864. "acquirements, it will be long before we see again.
"The *Mitis Sapientia Lælii* was never before better
"illustrated, and it was delightful to see so thorough
"a disgust for bigotry, in every shape, in Religion,
"Science, and Philosophy, coupled with such
"expansive charity and benevolence. A more
"charming character it never fell to my lot to
"associate with."

Barton,
June 5th, 1864.

My Dear Katharine,

Dear Charles and Mary's visit was a very
great pleasure to me: and most heartily sorry I
was when it came to an end. I enjoyed much
excellent talk with them, especially with him, and
though I was very sorry that the weather was
so odiously bad during their stay, yet it was a
satisfaction, that they saw this place in all its
beauty of blossom; for I do not think I ever saw
the trees and shrubs flower more richly. The way
in which the Hawthorns have been loaded with
blossom, is something really astonishing; and the
blossoming of the Lilacs, Laburnums, Horse-
chesnuts, Judas-trees, and Wisterias, has been
most profuse and beautiful. There is one Wisteria
in the Arboretum which has spread from the wall
over which it was trained, to a Fir tree that grew
at the back, and has climbed far up it, and is now
hanging beautiful wreaths of flowers from the
branches of the Fir tree; these higher shoots

flowering after the blossoms on the wall have
passed off.

Fanny has been most diligent and successful in hunting for birds' nests, and in watching the hatching and growth of the little ones. There are several that we have been watching daily. The green woodpeckers have been particularly busy and clamorous this season : their *goblin laugh* has been almost incessant. By the way, talking of goblins, have you read the "Diary of a Lady of Quality?" edited by Mr. Hayward. There are several good ghost stories in it.

I have got the concluding number of Species Filicum, and have arranged the Gymnogrammes and Nothochlaenas of my collection according to it.

Sir William Hooker is certainly a marvellously diligent and energetic man ; do you observe that, even in finishing this great undertaking of the Species Filicum, he announces another ? I should think this same Species Filicum must have been one of the most laborious works of our time ; much more so than its mere bulk would lead any one to suppose; much more so than many much larger and more showy books ; the number of specimens examined and compared must have been so enormous, and the difficulty of drawing lines between the endlessly varying forms often so very great. I consider the book a most valuable one as to species and their geographical distribution, but I am not satisfied to follow it as to genera and tribes.

1864. I hope you enjoyed your visit to Paris, and that dear little Rosamond did so too. I remember that, when we were there in '57, at much the same time of year, some of the gardens were in great beauty with a variety of flowering trees; the Judas trees particularly fine; and that was the only time I ever saw the Paulonia in flower, though I have heard it sometimes flowers in the South of England. It is a fine tree, strangely like the Catalpa in leaf, and in the general appearance (colour excepted) of the flowers, but belonging to a different family, Scrophularinæ,—the fruit I fancy, being different.

Clement is with us now :—a very fine lad, of good natural abilities, but by no means addicted to study as his brother Cecil is.

We have a Volunteer camp at Barton this week, in a large green field opposite to the Shrub wood, and Col. McMurdo is coming to inspect them on Thursday.

When I began this letter yesterday, I did not recollect that it was your birthday. I most heartily and earnestly wish you many happy returns of the day.

Much love to your husband and your dear children.

Ever your very affectionate Brother,
CHARLES J. F. BUNBURY.

JOURNAL.

[In the autumn the Charles Bunburys went to 1864. Bath, to attend the meetings of the British Association, where Charles Lyell was President; they then visited Chepstow and Tintern Abbey, and paid visits to the Bruces in South Wales, and to his brother Henry at Abergwynant in North Wales. They were accompanied by General Mc Murdo's eldest daughter Katharine, now Mrs. Ambrose].

_____

48, Eaton Place, London,
October 15th,

Read for the second time Charles Lyell's presidential Address to the British Association at Bath. Katharine and two of her children with Mary Lyell came to luncheon with us—also William Napier. Little Harry arrived from school.

_____

October 16th.

We (Fanny and Cissy) went with the children (Emmy and Harry) to the Zoological Gardens, and met Harry Lyell and all his children. Afterwards visited Katharine.

_____

October 18th.

Called on Bessy Arran and the Richard Napiers, and saw them both. We dined with the Henry Lyells.

_____

October 20th.

Mrs. Richard Napier came to luncheon with us, and Emily Napier arrived at the same time.

October 21st.

Lady Napier (Sir George's) came to luncheon with us.

October 22nd.

Dear Cissy and Emmy left us.

October 23rd.

Went to morning service at Vere Street Chapel, and had luncheon at the Lyells. Called on the Richard Napiers and had a long talk with them.

October 25th.

Visited Edward, just returned from his foreign tour : also Lady Bell. Edward dined with us.

October 26th.

William Napier came to us in the evening.

October 27th.

William Napier left us. Read Arnold's fourth lecture on "Military History:" admirable, especially what he says of reducing of towns by famine.

Dinner party : Henry Bruce, McMurdos, Mary Lyell and Edward ; Frederick Freemans came to us in the evening.

We went to Sandhurst to the William Napiers. I had a pleasant walk with William.

October 29th.

We returned from Sandhurst. The Henry Lyells dined with us.

October 30th.

Visit from Charles Lyell, and Mr. Twistleton —very pleasant.

October 31st.

The sad news of Lord Bristol's* death. We went down to Brighton.

November 3rd. Brighton.

Wrote my political letter to Maitland Wilson. Had luncheon with Lady Louisa Kerr. Saw Mr. Rickards.

November 4th. Brighton.

A most beautiful day. Took a walk with Mr. Rickards, who was very agreeable. Saw Sally and her daughters.

November 5th.

We returned to Eaton Place, and dined with Charles and Mary Lyell.

November 7th.

Heard from Maitland Wilson. Wrote on political business to him and to Scott.

* On the death of Lord Bristol, a new member for Suffolk was called for on his Son being removed to the Upper House.

1864.   Went with Fanny to the British Museum.   Frederick and Augusta Freeman dined with us.

November 8th.

Read the first three chapters of Creasy on the "English Constitution":—Good observations on the British or Celtic element in our nation ; curious list of Celtic words in our language, all relating to domestic, or menial, or feminine occupations.

The McMurdos dined with us.

November 9th.

Geological Society.   First meeting of the season. Dr. Duncan on Fossil Corals of Jamaica, on the several subdivisions of the Cretaceous system in the North of Ireland, compared with those of England, France and Belgium.   It appears, that the Irish Cretaceous beds consist of two main divisions, the upper (including the noted white limestone, which immediately underlies the basalt of Antrim) corresponding to the uppermost part of our white chalk, and perhaps in part to the Maestricht beds ; the lower to the Chalk Marl and Upper Greensand ; the whole of the Lower Cretaceous series, also the Lower White Chalk, are, it appears, unrepresented in Ireland.

November 10th.

We returned home to Barton.   All safe and right at home, thank God.   Read on our Railway journey the *Quarterly Review* on Dr. Newman's

" Apologia :" interesting in *so far* as it is made up of extracts from the book itself.

1864.

November 11th.

Received a letter from Lord Augustus Hervey, asking for my vote. Read chapters 4th, 5th and 6th of Creasy on the Constitution. In the chapter 4, treating of the Anglo-Saxon period, he points out the error of those who have represented the Anglo-Saxon institutions as a definite and settled system, which is far from being the case. The 5th and 6th chapters relate to the Norman Conquest. He shows reason for believing that the population of England, at the end of the 12th century, did not exceed at the utmost two millions.

November 12th.

Read Chapters 7th and 8th of Creasy. The 7th treats of the Feudal system in general; the 8th, its introduction into England by William the Conqueror, and of the modifications which he established in it, tending to increase the power of the Crowns and to diminish that of the great Feudatories in comparison of that of the corresponding class in France.

Read the first three chapters of " Cicero de Senectute" in the fine Paris edition which belonged to Mr. Fox.

November 13th.

Read part of the article on Noah (by Mr. Perowne in Smith's Dictionary of the Bible). The subjects

1864. of the Flood and the Ark are sensibly treated. The author is of opinion that the Deluge was partial, *not* universal; that the mountain on which the Ark rested was *not* that now called Ararat, but probably some part of the range bounding the plains of Assyria, and that the statement about Noah introducing "all kinds" of animals into the Ark, is not to be understood in anything near a literal sense. Read also great part of the article, in the same Dictionary, on the Physical Geography of Palestine:—excellent.

———

November 14th.

Read chapter 9th of Creasy on the English Constitution in which he treats of the condition of the "villeins" and the rest of the working classes of England, from the Conquest to the time of John; and the 10th chapter in which he gives an excellent account of the rising of the Barons against the tyranny of John, and the steps which led to the enactment of the Great Charter.

———

November 15th.

Visits from Maitland Wilson, John Phillips, Mrs. Wilson and Mr. and Lady Mary Phipps.

LETTER.

Barton.
November 16th, 1864.

My Dear Katharine,

Do you remember that I was sceptical as 1864. to the romantic story told in the *Edinburgh Review* about Bernard de Jussieu and the Cedar of Lebanon? Well, I have made out all about it from the article (a very interesting one) on Bernard de Jussieu in the *Biographie Universelle*. It appears that Bernard de Jussieu never went to Palestine at all, but he did bring two young plants of the Cedar *in his hat* to Paris,—*not* from the Lebanon, but from *England* : a much less arduous enterprise. The article, published in 1818, says that *one* of these Cedars is still living in the Jardin des Plantes :—this must be the one that we have seen.

The article in the *Biographie Universelle* is well worth reading ; he (Bernard de Jussieu) seems to have been a charming character.

I was very sorry not to see you before we left London, but I was confined to the house by a heavy cold on Tuesday, the day I had intended to make an expedition to Primrosia. I have quite got rid of my cold since we came home, and am very happy to be here. The lawn and park have been well-refreshed by dews and fogs, and now look English again, not African as they did in August. But

1864. there has not been rain enough to have any visible
effect on the ponds and wells, and the poor people
in some parts of the parish are suffering serious
inconvenience from the scarcity of water. Otherwise all seems to be doing well at present.

Mr. and Mrs. Bowyer arrived yesterday : you
know them I think,—very pleasant people.

The Ferns in our new Fern house are looking
well (except the Gleichenia) ; and we have got
some new ones from Wales, but they are still in a
doubtful state.

With much love to your dear children, and to
your husband,

I am ever,

Your very affectionate Brother,

CHARLES J. F. BUNBURY.

JOURNAL.

November 16th.

Mr. and Mrs. Matthew Arnold arrived.

November 17th.

Very wet and stormy.   Fanny went to Mildenhall
with Mr. Arnold* and Mr. Bowyer† to inspect the
schools.

I showed my Grandfather's drawings to Mrs.
Bowyer and to Mrs. Arnold.

---

* At that time School Inspector.

† Brother of Sir George Bowyer, and Inspector of Workhouse Schools.

November 18th.

Very blustery. We walked round the grounds
and visited the schools with Mr. and Mrs. Arnold.
Lady Caroline, Lister Kaye, Sir George Young,
Mr. and Mrs. Abraham and Miss Bethel arrived.
Lady Cullum dined with us.

November 19th.

Mr. and Mrs. Arnold went away, both very
pleasant.

The Maitland Wilsons and Miss Waddington
came to dinner.

LETTER.

Barton,
November 22nd, 1864.

My dear Joanna,

I ought to have written long before this
to thank you for your kind letter of the 4th of
October, which I was very glad to receive, though
I am afraid you will hardly believe this when I
have been so dilatory in replying to it.

What a tremendous storm and inundation you
have had at Florence ! It must have been interesting though awful. I read a startling account
of it in the *Athenæum*, written, I fancy, by Mrs.
Trollope.

Our party here which broke up yesterday has
been a very pleasant one. We have had staying
with us, Mr. and Mrs. Bowyer whom you may
remember ; Mr. Matthew Arnold and his wife,

1864. both very pleasant; Mr. and Mrs. Abraham and
Miss Bethel; Sir George Young and his brother
Edward, both of them remarkably pleasing young
men, intelligent, well informed, and thoroughly
gentlemanlike. Mr. Arnold is full of knowledge,
animation and vivacity, exceedingly amusing and
agreeable, and his wife is very interesting.

Altogether they made a very agreeable party.
Now we are alone for a little while, and I hope
to get some reading done, as well as business.

Since we came from London I have re-read
" Cicero de Senectute," an old favourite of mine,
and I am now engaged both on Merivale's " Roman
Empire," and Creasy on the " British Constitution." I am also reading to Fanny in the evenings,
Babbage's " Passages from the Life of a Philosopher;"
much of it is very entertaining, but I skip the details
relating to the Calculating Machine, which we
should neither of us understand. Mr. Eddis is
coming here at the beginning of next week to
stay some days with us, and to make for me a
drawing of Fanny in the same style as that beautiful
one of Susan. Fanny's letters will have told you all
about our interesting tour in Wales after the
scientific meeting at Bath. She was charmed as I
expected she would be, with Chepstow and Tintern
Abbey, and I who had seen them twice before (the
last time was 32 years ago) was not less charmed.

Duffryn, Henry Bruce's place, is in a country
naturally very pretty, but spoilt by collieries and
forges and furnaces. Abergwynant I thought as
lovely as ever, and though I was not quite well I

very much enjoyed our stay there with dear Cissy
and Henry and the children. Kate McMurdo, who
you know was our companion in all that tour, is a
very nice girl and very clever.

William Napier is appointed to the command of a
brigade at Dublin; he would have had to
give up Sandhurst at any rate early next year;
and this appointment is a very good thing, as it
saves him from the risk of being left unemployed
and idle, which he would hate.

I do not wonder that you were again charmed
with the beauty of Florence on returning to it. I
hope you will find it agree with you. We are both
very well I am thankful to say.

With much love to dear Susan,

I am ever your very affectionate Brother,

CHARLES J. F. BUNBURY.

JOURNAL.

November 23rd.

Read the last three chapters of "Cicero de
Senectute." It is a beautiful treatise : the latter
chapters particularly so. Cicero (like a practical
advocate as well as a philosopher) certainly makes
the most of his cause; he takes Old Age in
its most favourable aspect, and throws aside or
glosses over some of its most frequent evils. Thus,
he denies that the memory is weakened by age;
and he scarcely alludes to the various infirmities
and painful maladies, to which the old are so
often subject.

Read chapter 25th of Merivale : — Cicero's last gallant struggle on behalf of the Republic : his Philippics, the renewal of the civil war, and the battles near Mutina, in which both the Consuls, Hertius and Pansa were slain. There is a melancholy pleasure in seeing Cicero's fine genius and pure character, shining out in their fullest lustre so near the close of his life, on the eve of his own fall and that of the Republic.

Read to Fanny in the evening three chapters of Babbage's "Autobiography." The account of his descent into the Crater of Vesuvius and that of his visit to a Coal mine, are very entertaining

November 25th.

We resumed the catalogue of our books.

November 26th.

Read the remainder of the 26th chapter of Merivale. It gives the history of that sad and miserable period from the formation of the Triumvirate to the final extinction of the Republic at Philippi. It gives one a feeling of pain, even at this distance of time to read of such a triumph of wickedness in the persons of the three Cæsarian leaders. No doubt we must believe that the success of the enemies of liberty must have been a part of the scheme of Providence, and therefore in the long run, good for mankind : but the chiefs on the Republican side were so infinitely better

men than those on the other, that we can hardly 1864.
help repining at the result.

November 28th.

Mr. Eddis arrived.

December 1st.

Fanny's portrait concluded. We took Mr. Eddis
to Hardwick, on his way to the station, and saw
Lady Cullum.

December 2nd.

Arranged some minerals. Went on with the
catalogue of books. Read chapters 1st, 2nd and
3rd of "Cicero de Officiis," in Mr. Fox's copy.

December 4th.

Read chapter 21st of S. Luke, in Alford's edition,
with his notes; also his notes on Matthew,
chapter 24th.

December 5th.

We went to Stowlangtoft and had a pleasant talk
with the Rickardses

LETTER.

Barton,
December 6th, 1864.

My Dear Katharine,

Many thanks for your letter of the 23rd. I have looked at the Gleichenia Herrmannii (or dichotoma) in my herbarium, and find that the specimens from Rio Grande, collected by Mr. Fox, certainly have a different look from many of the Indian ones, being much less luxuriant, more contracted and pinched-looking—with something of a pauperized look; I suppose they grew in an unfavourable soil or situation; but the name was confirmed by Sir William Hooker, and they all have the distinguishing character he points out. It is not easy to describe this character in words, but I could show it you in a moment on the specimens themselves: it is a pair of outside pinnae, supernumeraries as it were, flanking each of the forkings of the stalk. Gleichenia glaucescens, which is very like Hermannii in general appearance, has not these. The specimens of Gleichenia Herrmannii from the Khasya mountains are very luxuriant. Hooker has sent me a great many from various parts of India, differing much in size, width of the pinnules and colour; it appears to be a most variable Fern. I hope you have seen Mr. Eddis's picture of Fanny, and that you like it. I am very much pleased with it. We found Mr. Eddis himself also very pleasant.

We have been reading Babbage's "Passages from 1863. the life of a Philosopher," have you read it ? Much of it is very entertaining, and it is everywhere very characteristic of the man. No book is more strongly marked of the individuality of the author.

With much love to your husband and children, and to Charles and Mary,

I am ever,

Your very affectionate Brother,

CHARLES J. F. BUNBURY.

I very much hope that you and Harry and your dear boys will come to us at Christmas. *Do* come.

JOURNAL.

December 7th.

Long talk with Mr. Darkin about repairs for Mildenhall Church.

December 8th.

Finished catalogue of books in dining room. Finished Matthew Arnold's little book on "Middle Class Education and the State." He begins with describing two establishments for education, which he visited when he was in France : one the Lycée at Toulouse, one of the great schools established and maintained by the Government:—the other, the "college" at Soréze, under the direction of the celebrated Lacordaire. After this, he proceeds to what is the main object of his treatise : to recommend for the Middle Classes of this country

1864. an organized system of education, assisted, directed, and controlled by the State. The question is a very extensive and difficult one, and I have no time to enter upon it here ; but Arnold supports his view with great ability, and in so agreeable a style and manner, as certainly to remove some of the repugnance which I felt to it at first.

December 9th.

Finished chapter 30th of Merivale's " Roman Empire."

December 13th.

Went on with the catalogue of my herbarium, read four chapters of " Cicero de Officiis." His observations on justice and injustice in relation to war, are good in themselves ; but it is rather diverting to find him representing the Romans as peculiarly and scrupulously observant of justice in their wars. We may perhaps say, that they were eminently observant of the *forms* of justice, and as remarkably regardless of its *substance.* We went on with the catalogue of books.

December 14th.

We arranged the print closet in the little library.

Friday, 16th December.

Read some of my Father's abstract of the Duke's dispatches.

December 19th.   1864.

Mr. and Mrs. Lombe and Julia arrived.

December 21st.

Lady Cullum and Miss Birch, Sir James Simpson and Mr. Tyrrell arrived : a large party at dinner.

December 22nd.

Harry and Katharine and their boys arrived, also Edward and the Abrahams. A very large dinner party,—pleasant evening—the ladies sang charmingly.

December 23rd.

All our guests went away except the Henry Lyells and Edward.

December 24th.

Showed part of the collection of minerals to Katharine and Leonard. A merry evening of games.

Saturday, 31st December.

The labourers' supper. Edward went away.

BOOKS READ IN 1864.

Froude's "History of England," vol. viii.—
A still shorter portion of Elizabeth's reign is contained in this second volume than in the first, no more than from January 1564 to February 1567.

1864. But two chapters are devoted entirely to the affairs
of Ireland; and the Scottish History which really
forms the main subject of the volume, is most
deeply interesting. It includes the critical part of
the reign of Mary Stuart:—that part in which,
as it seemed, she had the game in her own
hands; and all things seemed to promise her a
prosperous and triumphant career; when she
seemed to have baffled and humbled Elizabeth,
secured the favour and support of Spain, and all
but crushed the Protestant Lords with her brother
at their head. At the beginning of 1566,
immediately before the conspiracy against Rizzio was
formed, the Queen of Scots seemed to have before
her a much more brilliant prospect than Elizabeth
had. It was happy for England that she gave
herself up to the guidance of the Italian, and
united the fierce spirits of the Scottish nobility
against him and against herself. Mr. Froude has
shown us, much more clearly than previous
historians, what Rizzio really was:—that he was
not a mere artist pet of the Queen's, but a thorough
Italian of the Medici School, crafty, unscrupulous
and selfishly ambitious. The whole story of
Rizzio's murder, the Queen's escape from Holyrood,
her triumph over Darnley, and his murder are
wonderfully fine pieces of historical narrative.
Though Froude takes the more unfavourable view
of Mary's character, his portrait of her appears to
me (as yet) much more brilliant than that of
Elizabeth's.

What a magnificent creature Mary Stuart was!—

what splendid gifts, both bodily and mental, she 1864. had been endowed with ! Such strength and courage and energy and talents, as well as beauty and grace and fascination,—it makes me feel really unhappy that she should have been without principle and without conscience. But that she should have been without them is not at all extraordinary, considering the court in which she had been brought up. The most *unaccountably* evil part of her character was the bitter and malignant hatred she conceived towards her half-brother Murray : whose character by the way comes out very finely in this volume of Froude.

Elizabeth appears to still less advantage in this portion of the work than in the preceding. It would be difficult for any one's conduct to be more utterly shabby than her's towards Murray and his associates in 1565.

The idea given us in this book of the state of Ireland is exceedingly curious. The native Irish in these times would seem to have been rather less civilized than the Maories of New Zealand at the present day.

There is an excellent review of these two volumes of Froude (by Dean Milman as I understand) in the *Quarterly* for October 1863. Kingsley's article in *Macmillan* (January '64) is an eloquent and earnest apology for Elizabeth.

———

Robertson's " History of Scotland."—Books 2—8. After reading Froude, my curiosity and interest

1864. having been much excited about Mary Stuart, I
was naturally led to the *classic* history of her reign—
namely Robertson's: and beginning (but not
exactly at the beginning) with her marriage to the
Dauphin of France, I finished the work. I must
confess that, at first, the style did not much please
me ; it did not appear to advantage immediately
after Froude, it seemed to me stiff and formal, too
much studied, too elaborately and uniformly
polished. But after a little while I became
accustomed and reconciled to it, I was led on by
the interest of the story, and read on with pleasure
to the end. Still I cannot but think that Lord
Brougham's praises of the style of the book (in his
"Men of Letters") are exaggerated. Sir James
Mackintosh's remarks in his Journal, agree exactly
with my taste ; he says—"I think the merit of
Robertson consists in a certain even and well-
"supported tenour of good sense and elegance.
"There is a formality and demureness in his
"manner ; his elegance has a primness and his
"dignity a stiffness which remind one of the
"politeness of an old maid of quality, standing on
"all her punctilios of propriety and prudery.
"These peculiarities are most conspicuous in his
"introductory book. As we advance, his singular
"power of interesting narrative prevails over every
"defect. His reflections are not uncommon : his
"views of character and society imply only sound
"sense."*

To this I must add that even in the latter parts 1864. and throughout the book, although I was much interested in the history, there still remained to a certain degree the same impression of a certain *full-dress* formality of style, and also (what is a greater fault) of too much uniformity, something too regularly and invariably polished and worked up in the composition. There is also to my taste, too great a frequency of rather trite and common-place moral reflections :—at least what seems so now ; in the middle of last century they may not have appeared quite so obvious. The style has, however, the great merit of perfect clearness ; we never have to read a sentence over a second time to make sure of its meaning, as is so often necessary in reading "Gibbon."

This history is in itself one of the most romantic imaginable :—absolutely crowded with remarkable characters and extraordinary events. Mary's own adventures are well-known as pre-eminently romantic ; and then there are the striking circumstances of Murray's murder ; the extraordinary escalade of Dunbarton Castle by Crawford, (extremely well-related by Robertson), the surprise of the Regent Lennox and his party in Stirling by Kirkaldy of Grange and their rescue by Erskine ; the downfall of Morton, the strange turns of fortune in the struggle which so long went on in the youth of James VI between his personal favourites and the principal nobles : the repeated attempts to seize the person of James, and at last, most strange and unintelligible of all, the mysterious Gowrie con-

1864. spiracy. Truly the Scots of those days were fierce and turbulent and difficult to govern : and those who held or sought high station had no easy or pleasant task. Most of those who took a prominent part in the public affairs of Scotland in Mary's reign or during the minority of James, came to a tragical end. Darnley murdered, Rizzio murdered, Bothwell dying mad after long imprisonment in a foreign country, the Regent Murray murdered, the Regent Lennox murdered, Morton beheaded, Archbishop Hamilton hanged, Kirkaldy of Grange hanged, Maitland of Lethington dying by his own hand, here is a tragical list. It is rather extraordinary that in such times, a man of the character of John Knox should have died peacefully in his bed.

Robertson expresses himself much more cautiously than Froude on the subject of Mary's guilt ; but it is tolerably evident that he believes her to have been guilty—certainly of adultery with Bothwell, and more or less of complicity in the murder of Darnley. He shows that feeling of interest in her character and fate, which it is hardly possible, even for those who judge her most severely, to resist ; but yet he by no means gives us so high an idea of her natural gifts and splendid qualities as does Froude. In Robertson, as in the ordinary historians and romancers, he sees Mary as a woman gifted with extraordinary powers of fascination : but Froude shows her gifted with courage, energy, talents, almost every brilliant endowment of body and mind—

" And wanting nothing but an honest heart."

H. B. Tristram. "The Great Sahara." 1 vol:— 1864.

This is an excellent book of Travels: written in a clear, pleasant, unaffected style, and full of really sound and satisfactory information. The author is an excellent and zealous naturalist: his notices of the animals which he met with, and his observations on the geology of the country, are copious and interesting; and one feels sure that what he says on these subjects may be thoroughly trusted. To botany he seems to have paid less attention, but he gives in one of his appendixes, chiefly on the authority of a French botanist, a copious list, very far more copious than I should have expected of the native plants. Ornithology seems to be his favourite study, and in the course of the work he gives much curious incidental information about the birds of the country, besides the more full and collected and systematic review of the subject in his appendix. It is particularly interesting to see that he met with many of our most familiar *summer* visitors abounding in the oases of the Sahara in winter; the Swallow, the Martin, the White Throat, &c. The descriptions of the physical geography and of the scenery, give very clear ideas of the wild, strange, desolate country. Tristram's theory is, that in a late geological age, a broad arm of the sea extended across Africa, from the Gulf of Cabes between Tunis and Tripoli, to the Atlantic, opposite the Canary Islands, and that the Sahara is an elevated and dried-up bed of the sea. He shows very strongly the probability of this opinion.

1864.   Forsyth's "Life of Cicero."   2 vols.—

This life of the great Roman orator is interesting
and agreeably written ; not, I think very profound
or very original in its views or in any way remarkably striking, but very pleasant reading.   The
author has a proper admiration for his subject,
without the blind idolatry of Middleton.   He does
not disguise or gloss over, nor attempt to defend
the many faults and weaknesses of Cicero: his
enormous vanity, his irresolution, vacillation and
political timidity; but he dwells strongly and with
much reason on his many and great merits.

Indeed, in spite of those faults and weaknesses,
Cicero appears to me by far the best man of his
time.   There is something very attractive and
loveable in his character, and his moral principles
were high, though in public life, through want of
firmness, he often fell lamentably short of them.
His private life seems to have been more pure than
most of the eminent men of his time, though I
cannot excuse him for divorcing his wife without
reason assigned after thirty years of wedlock.

——— —

Charles Kingsley.   "The Roman and the
Teuton," (lectures delivered before the
University of Cambridge).—

This is Kingsley's first course of lectures to his
Modern History Class at Cambridge.   The subject,
the Conquest of the Roman Empire by the
successive invasions of Teutonic Tribes, Goths and
others ; the overthrow of the old order of things

and rise of a new, down to the formation of a new 1864. Empire under Charlemagne. The lectures are very interesting and extremely characteristic of the author : I recognize him in almost every sentence : often in reading them, I can fancy I hear him speaking. They are very spirited, full of fire and energy, very imaginative, often very eloquent, sometimes very fanciful, and sometimes very *odd*. They are pervaded throughout, moreover, by that strong religious sense, which is as characteristic of Kingsley as of Arnold ; that conviction (in which I think him perfectly right) that all the events of history and all the actions of man, as well as the laws of nature, are constantly directed and regulated by the providence of God.

The first lecture in which he takes a general view of the Teutonic nations, and illustrates his idea of their condition by a myth or "sage" of his own, is striking and very peculiar, and the idea which he here seeks to illustrate, that the Teutons at the time of their struggles with Rome, were huge overgrown children, children in mind and giants in body, recurs again and again through the book ; it is in fact the leading idea of his account of them. The second lecture, describing the state of Rome just before its final overthrow, the "dying Empire" is very powerful : perhaps here and there a little exaggerated. The two lectures, entitled "The Clergy and the Heathen," and "The Monk, a Civilizer" are remarkably interesting : perhaps the best in the book. In the one on the "Lombard Laws," Kingsley points out in what respects the

1864. influence of the Priesthood had an injurious effect upon the laws: how the prejudices connected with their celibacy and other habits, observances and principles of their profession, led them to favour the use of torture and of cruel and degrading punishments, and to run into all the atrocities of the Witch persecutions. He here notices, what I remember his mentioning to me at Barton in '61, it then struck me as new,—that corporal punishment of any free person, however young, was quite repugnant to the notions and habits of the Teutons, and that the Monks were the first who re-introduced the practice of flogging in schools. The last lecture, "The Strategy of Providence," appears to me a little fanciful.

The inaugural lecture which is placed at the beginning of this volume, is not directly connected with the subject of the other lectures. It was published separately, soon after it was delivered, and I read it at that time. Its subject, "The Limits of Exact Science as applied to History," is a very difficult one, and though I agree in the main with Kingsley's view of the question (so far as I understand it) as opposed to Buckle's, I cannot say that I think his treatment of it altogether clear or satisfactory.

———

Colonel Charras. "Histoire de la Campagne de Waterloo." 1 vol.—

A very clear and well written narrative and military analysis of the Waterloo Campaign: fair and candid, and even liberal towards the Allies, to a 1864. degree that really surprises me in a French narrative. It is certainly a great contrast to the way in which French writers generally have treated this campaign. It is to be observed that (unlike most of his countrymen who have written on this subject) Charras is by no means an Imperialist: he is evidently a zealous Republican: and that, far from acknowledging Napoleon as an infallible authority, he condemns strongly and unsparingly his political measures, points out without reserve what he considers his military errors in the conduct of the campaign, and exposes the misstatements in his account of it. He shows conclusively the untrustworthiness of the Emperor's statements, dictated at St. Helena, as to the Waterloo campaign: that they were in fact explanations invented *après coup*, in order to throw off from himself all the blame of the failure; the character that Charras draws of the Duke of Wellington appears to me remarkably just. He does justice too, to the firmness and enduring courage of the British Troops at Waterloo, especially in their resistance to that tremendous *storm* of cavalry which Ney directed against their centre; of course there may be and will be disputes as to the accuracy of his statements, as to particular details of the great battle; it is not likely that there ever will be entire agreement or absolute certainty as to these: but on the whole I cannot but believe that his narrative is essentially trustworthy.

According to Charras, Napoleon's conduct all

1864. through this campaign, was marked by a peculiar
degree of slowness, hesitation and irresolution, a
remarkable want of that activity and decision so
conspicuous in his earlier wars. He was in fact,
according to this account, enfeebled both in mind
and body, and had lost much of that fiery energy
and indefatigable activity which had contributed so
much to his wonderful successes. " Napoleon was
old before his time." His health was bad, his
bodily strength and activity were much broken, he
could no longer dispense with sleep, nor endure the
fatigue of rapid journeys, and long days spent on
horseback, as he used to do ; and though his mind
had still the same facility and richness and force of
ideas, he had lost much of his former rapidity of
decision and firmness of purpose. So says Charras,
and since I read this book, I have found a curious
confirmation of his opinion in " Soult's Conversations with Sir William Napier, in 1838," with which
Charras could not have been acquainted. Speaking
of the Waterloo campaign, Soult said, " But the
" Emperor seemed at times to be changed ; there
" were moments when his genius and activity
" seemed as powerful and as fresh as ever ; at other
" moments he seemed apathetic. For example, he
" fought the battle of Waterloo without having
" himself examined the enemy's position. He
" trusted to General Haxo's report. In former
" times he would have examined and re-examined it
" in person."*

On the other hand, Charras blames the Duke

* Life of Sir William Napier.   Vol. 1, p. 505.

(and seemingly with reason), for dispersing his army 1864. too much in the beginning of the campaign ; for slowness in concentrating it even after the attack on the Prussians had begun ; and above all, for keeping even to the last a strong detachment of not less than sixteen thousand men idle and unemployed at Hal far away on his right, several leagues from the field of battle. He would seem to have been haunted all along by an unfounded apprehension that the French would attempt to turn his right, and get to Brussels before him. Charras shows that such a movement under the circumstances would have been contrary to the maxims of the science of war, and especially contrary to the practice of Napoleon.

Napoleon's plan of campaign seems as far as I can judge, to have been both grand and skilful in its conception. Knowing that the two armies, the Prussian and the British were unused to co-operate, were commanded by two chiefs of extremely different characters ; and were widely separated in their cantonments, he thought that he could fall upon them separately, and crush first the one and then the other, before they could give any mutual assistance. Charras thinks that this scheme might have been successful if carried out with the vigour and alacrity of his earlier days. At any rate he did attack the Prussians separately at Ligny, and was to a certain degree victorious over them ; while Ney, though unable to beat the English at Quatre-bras, succeeded in preventing them from helping the Prussians. But the Emperor's great error seems to

1864. have been in supposing that the Prussians had been so thoroughly beaten at Ligny, as to be incapable of affording any help to their allies.

Accordingly, as Charras observes, his plan of attack at Waterloo was based on the assumption that the Prussians could take no part in the battle. Charras also blames him severely for sending his last reserve "the Imperial Guard," into action, instead of using them to secure his retreat, when it became known that the Prussian army was in force on his flank. But probably he felt that he was playing in every way a desperate game, and he played it desperately. The Prussian commander certainly showed great energy and ability in determining upon, and executing the movements which brought them on the right flank of the French on the 18th ; and this so soon after the terrible and sanguinary battle they had fought at Ligny. Pertz tells me that the merit of this is due to Gneisenau, who decided on the movement and gave the orders ; for Blucher, having been thrown down and ridden over by the French cavalry was in a state of insensibility or nearly so for many hours.

As to the events which followed the battle of Waterloo, Charras fully confirms what I had learnt from Villemain ; that Napoleon was lost from the moment he returned to Paris ; that his chances were over when he quitted his army.

"Life of Sir William Napier." 2 vols.—

This is to me a singularly interesting book, undoubtedly it cannot be expected to have quite the same degree of interest for strangers that it has for me ; but I think no one who has anything of nobleness in his composition can fail to be interested in the grand and splendid character that is here exhibited. The book too, I think, is well done as a biography : the principal part of it is, as it ought to be, made up of the letters and other manuscripts of Sir William, but the connecting thread of narrative, and the remarks and explanations, are in very good taste and feeling, modest, discreet and judicious. This connecting composition is principally if not entirely the work of Patrick Macdougal ; the final selection of letters and other documents for insertion was made by Henry Bruce, whose name appears on the title page as editor. Both have done their work very well. The biography undoubtedly is, as biographies according to rule and precedent ought to be, a picture giving the most favourable likeness ; it is a portrait, not indeed without shadow, but with the shadows skilfully softened. The harsher traits are just delicately indicated, and at the same time fairly explained and accounted for. In truth, I think this book will do a great deal for Sir William's memory ; for many may have heard of his violence and irritability of temper, without being aware of the dreadful sufferings from ill health and wounds which contributed so much to occasion those infirmities. I suppose few men ever suffered more pain or for a greater length of time. I daresay his temper was originally bad—indeed I have heard that it was so, but certainly its faults were

1864. prodigiously increased, and at the same time in great measure excused by his sufferings. Those faults of temper were annoying enough in his lifetime to those who had anything to do with him : but in recollection and in reading of him, they shrink almost to nothing in comparison with the heroic greatness of his character and the splendour of his intellect.

The letters written by him during the Peninsula campaigns and the extracts from a manuscript memoir of his services in that war are peculiarly interesting. This memoir of his services, I remember, he gave me to read on one of my visits to him at Freshford, soon after it was written, I fancy ; and I have a lively recollection of some of the most interesting parts of it, which are now published in this book. One is very much struck in reading these letters, with the remarkable degree of sensibility and tenderness of feeling which they show : an extreme acuteness of feeling, an almost excessive sensitiveness to the sufferings of others which are very remarkable in combination with such fierce daring. His feelings of all kinds, indeed, whether of tenderness or of fierceness, were alike vehement and tending to excess. It appears that the intensity of his feeling for the sufferings he witnessed, his distress on account of the loss of friends in battle, and his indignation at the atrocities of war, excited in him at times an absolute dislike to the military profession, and actually gave him a wish to retire from the service. This is surprising enough to those who knew in later times his

attachment to his profession, and how he delighted 1864.
in talking of that war.

Some of his peculiar opinions and sentiments appeared strongly marked very early in his correspondence: his liking for the French, his enthusiastic admiration of Napoleon, his vehement prejudices against the English ministers, his readiness to believe any evil of those whom he disliked (see for instance the letter on the death of Perceval, and that on the attempt to assassinate the Duke of Wellington) and equal readiness to ascribe every good quality to those for whom he had a liking.

A remarkable example of the care, the sagacity and the power of intellect with which, while quite a young man, he studied and judged of military questions, is seen in his remarks on Napoleon's Russian campaign in the letter dated September 27th, 1812.

The biographer has shown, I think, much discretion and good taste in dealing with the numerous disputes and angry controversies in which Sir William was so unfortunately ready to engage. I approve especially of his remarks on the disputes with Mr. Dudley Perceval; on all occasions, indeed, his reflections appear to me to show good feeling and good sense.

The letter from Sir William to Lady Hester Stanhope (dated March, 1839), in which he gives an account of his own career, his position and his family, is a peculiarly beautiful and interesting one; perhaps the most interesting in the book.

I had never heard of it before. I am told that it

1864. never reached Lady Hester : that she was dead
before it arrived, and it was sent back. Sir
William's memorandum (vol. i., p. 28) of his acquaintance with Pitt, is singularly interesting and
curious : it has given me quite a new idea of the
great statesman in his hours of relaxation ; indeed,
a much more distinct idea of his real character, as a
man, than I received from the whole of Lord
Stanhope's four volumes.

"Life of William H. Prescott," by G. Ticknor. —
An extremely agreeable book.   Mr. Prescott's
own letters are delightful, especially those written
from Europe in 1850.   They quite justify the
accounts of his singularly attractive and lovable
character, which I have received from all that knew
him ; and they increase my regret that, owing to
unavoidable circumstances, I should have lost the
opportunity of becoming acquainted with him when
he was in England in that year.   His character
seems, indeed, to have been one of the most
charming that we can conceive, a singular sweetness
and delicacy of feeling and warmth of affection,
united with the remarkable energy which was
required to overcome such physical disadvantages
as he laboured under.   It is certainly remarkable
how much he achieved under such impediments as
those of blindness and frequent ill health.   He was
not, it is true, totally blind, but so nearly so, that it
is really astonishing how he contrived not merely
to write so many and such good books, but to go

through such an amount of study and research as 1864.
was necessary to supply the materials for them.
He lost entirely the sight of one eye by a singular
accident when he was very young: and not long
after the other eye suffered so severely from a
violent inflammation (of a rheumatic nature) that he
was for a long time deprived of the use of it, and it
continued very weak to the end of his life. He
was always obliged to be very cautious in the use of
it, and was often disabled from using it at all for
considerable periods. He likewise continued, all
his life, to be liable to severe suffering from
rheumatic attacks; under these circumstances, we
can hardly admire too much the courage and
perseverance which enabled him to accomplish so
much.

Arnold's " History of the Latin-Roman Common-
wealth." 2 vols.—
These two volumes contain the history of Rome
from the end of the second Punic war to the death
of Augustus; written originally by Arnold for the
Encyclopædia Metropolitana, in which it was published in 1823; and collected and re-published
in these volumes after his death, as a sequel to his
Roman history. This is certainly a less mature
work, and less perfect as a model of historical
composition than that history of the earlier times of
the Republic, which was so unfortunately cut short
by his death; besides that, the plan of it did not
admit of so full and copious a treatment of the

1864. matter; but still it is a noble and a valuable work. The moral greatness of Arnold's mind shines through it all, and gives a peculiar dignity and impressiveness to the narrative; and while one reads this book with deep interest, one regrets more than ever that he should not have lived to treat at full length and in the full maturity of his judgment and knowledge, a period so much more interesting than the semi-mythical times before the Punic war. The style of this book is clear and dignified, but stiffer and less animated, it appears to me, than that of his later History.

But the especially striking characteristic of the book is, as I said before, the lofty moral tone which pervades it throughout; a morality, strict, austere, perhaps sometimes erring on the side of sternness; but always pure and noble. No splendour of achievements, nor grandeur of intellectual character can soften the severity of his judgments. His impartial severity is well shown in his treatment of the character of Julius Cæsar on the one hand and of Cæsar's murderers on the other. His condemnation of Cæsar is unqualified and unsparing; it would hardly be too much to say that his feeling towards Cæsar amounts to abhorrence. He denounces again and again with strong and just indignation, the great Dictator's utter want of political or moral principle, his selfish ambition, the recklessness with which he sacrificed whole tribes and nations to the attainment of his own personal objects. He goes so far as to say that probably no man ever caused a greater amount of human misery with less provocation or temptation. But on the other hand, he has little sympathy with the assassins of Cæsar; he condemns their act with the same dignified severity, and shows how unjustifiable was the conduct of even the best of them.

Perhaps Arnold hardly makes sufficient allowance for the temptations of Cæsar's position; the temptations (I mean) held out to a man naturally so ambitious, by the disturbed, confused and tumultuous condition of the Republic. He shows a disposition to believe that the disorder of the Republic was not irremediable; that the old constitutional freedom might have been preserved, and the domination of any one man averted. I confess that even from his own narrative I am not satisfied of this. Nor can I entirely go along with him in his admiration of the character of Pompey, who is evidently his favourite hero. Pompey's private character was indeed estimable (though he does not appear to have possessed equally with Cæsar, the gift of attaching personal friends), and he showed humanity as well as ability in his operations against the pirates. But his behaviour to Cicero was most shabby; his conduct throughout the Triumvirate was at once unprincipled and weak, and decidedly, I think, prepared the way for all the subsequent calamities of the state and his own. Nor can I see any good reason to doubt that if victorious in the civil war he would have gratified the vengeance and rapacity of his followers by extensive proscriptions.

I had feared that Arnold would have been very

1864. severe upon the weaknesses of Cicero; but he judges him very fairly and kindly, with a full appreciation of the merits and due indulgence to the faults of that most amiable of great men.

Coming on to the time when Augustus had over-thrown all his rivals and stood supreme, Arnold clearly shows how easily the Republican constitution of Rome subsided into a practical despotism, by the simple process of accumulating in the hands of one man, all the offices and powers which were before divided.

It cannot help striking anyone who reads the history of the Roman Republic with any care, how very arbitrary and enormous were the powers entrusted severally to each of the great officers—the Consuls, Censors and Tribunes—and that nothing but their mutual opposition prevented them from becoming actually tyrannical. So when one and the same man became at once Consul, Censor, and Tribune, without any colleague, or with merely a nominal one, the Government slid at once into a practically absolute monarchy, without any ostensible or avowed change of the constitution.

In reading the chapter in this book on the reign of Augustus, I was again forcibly struck, as I have often been before, with the remarkable analogy between the Empire of Rome under Augustus, and the Empire of France under Napoleon the Third. I think Dr. Arnold would have been much struck with this if he had lived to the present day. It is not often that one finds historical parallels so close as that between the nephew of Julius Cæsar and the nephew of the great Napoleon. I do not indeed 1864. mean to say that Louis Napoleon is as great a villain as Augustus the Triumvir, but nevertheless the likeness is remarkable. So, too, there is a striking analogy between the Governments they have respectively established; in each case a practical despotism veiled under constitutional forms. But will the parallel be continued? I doubt very much indeed whether the elements of revolution in France are exhausted, as they were at Rome at the time of Augustus.

Rome under the Emperors was a burnt-out volcano; France, I imagine, is only a slumbering one.

Charles Merivale's "History of Rome under the Empire." Vols. i. and ii.—

These first two volumes embrace the same period of history which occupies the greater part of Arnold's two volumes,—from the Consulship of Cicero (A.D. 691) to the death of Julius Cæsar (A.D. 710) (B.C. 44), 19 years of wonderful excitement, agitation and interest. Merivale's book is a good one, but one misses in it that stamp of a master mind which impresses one so strongly in Arnold's writings. Merivale is clear, painstaking, and evidently conscientious, but somewhat heavy; his style wants simplicity; it is too evidently laboured, too evidently straining after effect : this is more particularly the case in his general reflections; his narrative is generally clear and good,

1864. especially in the military operations. The battle of Pharsalia is particularly well related.

It is curious and instructive to observe the wide difference, and indeed strong opposition, between Arnold's and Merivale's views of the characters and actions of men at that time ; while both draw their conclusions from the same facts. Pompey is Arnold's hero ; Cæsar is Merivale's. Both writers of course had the same material, the same sources of information ; both relate the same events, with little difference as to matters of fact ; but the inferences they draw from these facts, as to the characters of the two great rivals, are on almost every point absolutely opposite. Merivale indeed disposes of some of the worst stories touching Cæsar's morals, by simply disbelieving them as mere scandal : and it is no doubt possible that he may be right. He also urges with force the reasons (and strong reasons undoubtedly) for believing that the old republican government of Rome was hopelessly decayed, and that its termination in monarchy under some form or other was inevitable. I have already said, that I am inclined to think Arnold mistaken on this point. It appears to me from all the information we have concerning those times, that the republican government was virtually dead ; existing merely in form and name, utterly ruined by violence and corruption ; that public spirit, public virtue, reverence for law, no longer existed to an extent sufficient to make its restoration possible. Even the best men of the time, even Cato and Cicero, confessed virtually and practically, though

not in words, that the peaceful administration of the 1864.
State according to the laws had become impracticable. They seem to feel that it was become
hopeless to oppose corruption and mob violence,
except by other bribes and other mobs. If Pompey
had been victorious in the civil war, and had
abstained (as with his moderate and cautious
character he very possibly might have abstained)
from seizing on the supreme power for himself, the
consequence would probably have been that he
would have been thrust aside by his own party;
and after a certain period of lawless and selfish
struggles, some other adventurer would have arisen
who would have played less worthily the part of
Cæsar. On the other hand, Merivale certainly
shows himself disposed to treat Cæsar with excessive
indulgence, while he studiously puts the most
unfavourable construction on all the acts of Pompey.

" Cicero de Senectute."—
The fourth time I have read this beautiful
treatise.

Poulett Scrope on "The Volcanos of Central
France." 1 volume, edition 2.—
This is in some sort a supplement or companion
to the same author's general work on Volcanos.
Like that book, it is clearly and agreeably written,
well arranged and very instructive. Much of it is
strictly descriptive, and would be more especially

1864. valuable as a guide to the geological traveller.
Scrope begins with a general description of the
elevated granitic region of Central France; on
which all the volcanos have broken out. He next
describes the tertiary fresh-water deposits of the
Limagne d'Auvergne, and others. Next, he
proceeds to describe in great detail the Puy de
Dôme and the numerous volcanic cones, craters and
lava currents in the same tract of country; then he
examines in the same manner the volcanic region of
the Mont Dore, then that of the Cantal, and
lastly, that of the Velay and Vivarais or the
departments of the Haute Loire and the Ardèche.
In the course of these descriptions we meet with
many interesting and important observations and
discussions of a more general nature, concerning
various points of volcanic geology. Such are the
remarks on the mode of formation of the Puy de
Dôme and other similiar trachytic mountains
without craters (p. 49) on the elaboration of trachyte
and basalt from granite and hornblende rock by
volcanic action (p. 76) on the passage of trachyte
and basalt into one another (p. 118 and elsewhere)
on the formation of volcanic conglomerates or
tuffs, by torrents of water rushing down the sides of
the volcanos during eruptions, and sweeping along
vast quantities of the loose ejected matters
(p. 120) on the mode of formation of those
peculiarly wide and shallow craters, which in the
Eifel, on the Rhine north of Coblentz, are called
Maare (p. 181, 182)—and above all, the excellent
disquisitions, in the last chapter particularly, on the

wearing down of rocks and excavation of valleys, 1864.
by the long continued action of the every-day
ordinary agents, rain in its various forms, frost and
running streams, such as are still everywhere at
work.

Scrope repeatedly and forcibly points out, what
we are apt to forget,—that currents of lava would
naturally, as a general rule, tend to run into the
valleys and lower levels of the country : and therefore, that, when we find them crowning high
plateaus and promontories, and capping outlying
bold hills, the fact necessarily implies an immense
amount of denudation. As to the causes of this
denudation, he contends as strenuously as Charles
Lyell himself, for referring all to the action of
existing causes, continued through indefinitely long
periods of time.

This book is illustrated by numerous and very
interesting views of the scenery of the volcanic
districts, which give a lively idea of the remarkable
forms and grouping of the mountains, and the
splendid ranges of columnar basalt which crest the
heights or border the valleys. Some of the views in
the Velay and Vivarais are especially striking

———

Charles Merivale's "History of Rome under the
Empire." Vol. iii.—

This volume brings down the history from the
death of Julius to the complete establishment of the
Empire under Augustus, from the year of Rome
710 to about year 731. With respect to the merit

1864. of the work itself, I may repeat what I said of the first two volumes, that it is evidently a work of honest conscientious labour, and very instructive. It is strange what a want of foresight and wisdom was shown by the conspirators against Julius Cæsar. They had organized nothing: they had not as it appears formed any kind of plan or system for following up this blow, and when the Dictator had fallen they knew not what to do next. They seemed to have expected that the Republic would restore itself spontaneously: that all would settle back into its old course as soon as the usurper was taken away; so completely had they miscalculated the feeling of the people, that within a very short time after their successful blow, they were obliged to hide themselves, and to escape secretly from the fury of the veterans and of the mob of Rome. Decimus seems to have been the only one of them who showed either military or political ability. Yet, for all this, the men who were on that side were so much better than those on the other, that one always feels an emotion of sorrow in thinking of their fate. The story of the Triumvirate and the proscriptions, the murder of Cicero, and the battle of Philippi is certainly one of the painful portions of history.

Nevertheless, one can see, that the overthrow of the republican party and the establishment of a moderate monarchical government, such as that of Augustus, was on the whole conducive to the general good of mankind. There was something

narrow, hard and exclusive in the spirit of the 1864.
aristocratic republican party,—the party of the
senate, to which Brutus and Cassius belonged. (I
do not, however, think it fair in Merivale to
designate this party as he does constantly by the
invidious title of "the oligarchy"). They were
unwilling to extend the benefits of Roman citizen-
ship, even to the Italians : much more would they
always have looked down on the provincials ; such
an amalgamation and fusion of all parts of the
Roman Empire, as entered into the scheme of
providence, could hardly have taken place under
their government. Merivale, whose imperialist
leanings are certainly very strong, shows satisfactorily in the last pages of this volume, the
advantage, and almost the necessity of the
establishment of the Empire.

LETTERS OMITTED.

(From Sir Charles Bunbury to his Brother Edward).

Mildenhall,
June 8th, 1853.

My Dear Edward,

Many thanks for your letter. I had 1853.
already seen in the newspaper the decision of the
Bury election committee ; and very sorry I am for
your disappointment, for I fear you are right in
thinking that this result nearly destroys your chance
of again coming in for Bury. It is very provoking,
and I do feel much for you, as you had made up

1853. your mind to adopt parliamentary life as your particular career, and it is most mortifying to find yourself thus excluded from a course of life which suited you, and in which you were likely to succeed. Still you have abundance of other resources, and it cannot be the same loss to you that it was to a man like Shafto Adair, for example, who had made political life his sole pursuit. Fanny sympathizes heartily with you, and is quite indignant at the decision of the committee. She is I am happy to say, very well, for her, and we are much enjoying the quiet of our home, and the delightful weather which has come at last. I am very sorry indeed that Emily is again so ill.

Ever your affectionate Brother,
C. J. F. BUNBURY.

It is strange that I should have forgotten to say anything about your very handsome donation of books to the Mildenhall Institute. The Committee were mightily pleased, and no wonder, for it is certainly one of the most liberal gifts the Institute has ever received. I suppose you have had a *regular* and *proper* letter of thanks on the subject? The books are capitally well chosen, and will I think be very popular; I am sure they ought to be.

48, Eaton Place,
December 1st, 1863.

My Dear Edward,

1863.   I received yesterday your letter from Smyrna and Athens, which has entertained and

interested me extremely; and I thank you very 1863.
much for it. We have been just a fortnight in
town, and have seen a good many pleasant people,
and dined out most days. To-morrow we intend to
return to Barton. From the 10th to the 18th or
19th of this month we expect to have again a good
houseful of company, and we both of us hope that
you will come to Barton during that time. We
shall have the Henry Bruces and Catty and George
Napier, the Frederick Greys, the Adairs, and Sir
George Young; your room shall be ready for you
and we shall be very glad to see you.

Very good accounts from Wales. We saw the
Richard Napiers the day before yesterday, and
found them remarkably well, all things considered.

The season has been generally a very unhealthy
one, and fevers of various kinds are very prevalent;
but Fanny and I are very well, I thank God.

I will not write more just now, as I hope to see
you soon.

Ever your affectionate Brother,
CHARLES J. F. BUNBURY.

MILDENHALL :
PRINTED BY S. R. SIMPSON, MILL STREET.